ALSO BY JAMES S. ALBUS

Peoples' Capitalism: The Economics of the Robot Revolution

Brains, Behavior, and Robotics

Intelligent Systems: Architecture, Design, and Control, with Alexander Meystel

*Engineering of Mind: An Introduction to the Science of
Intelligent Systems*, with Alexander Meystel

Intelligent Vehicle Systems: A 4D/RCS Approach,
Raj Madhavan, Elena Messina, James Albus, Eds.

PATH *to a* BETTER WORLD

||||||||||||||||||||||||||||||||||||||||||➤

*A Plan for Prosperity,
Opportunity, and Economic Justice*

James S. Albus

iUniverse, Inc.
Bloomington

Path to a Better World
A Plan for Prosperity, Opportunity, and Economic Justice

iUniverse books may be ordered through booksellers or by contacting:

iUniverse
1663 Liberty Drive
Bloomington, IN 47403
www.iuniverse.com
1-800-Authors (1-800-288-4677)

ISBN: 978-1-4620-3532-8 (sc)
ISBN: 978-1-4620-3533-5 (hc)
ISBN: 978-1-4620-3534-2 (e)

Library of Congress Control Number: 2011913275

Printed in the United States of America

iUniverse rev. date: 11/16/2011

To my wife, Cheryl

PREFACE

About the Author

As my name is not a household word outside of my field of intelligent systems research, it seems appropriate to provide some background as to who I am. From an early age I have been fascinated with problems of how things work and interested in how to make things work. High school physics introduced me to how force and mass interact to produce motion and transform energy into different forms. Geometry and trigonometry taught me how the structure of the world can be described and analyzed. The Radio Amateur's Handbook provided me with a wealth of knowledge about electric circuits and vacuum tubes. As a summer student in college, I worked at the Naval Research Laboratory (NRL) in Washington, DC, in a sounding rocket program that was later incorporated into "Project Vanguard" with the intent to produce the world's first artificial earth satellite. My assignment was to design the antenna and feed network for that satellite. Unfortunately, the Russians launched "Sputnik" before the first US launch attempt, and the first Vanguard rocket blew up on the launch pad. As a result, the Von Braun team in the NASA Huntsville space center placed the first American satellite in orbit. However, my satellite finally made it into orbit where it will remain for about two hundred and fifty years. A spare model hung in the Smithsonian Air and Space Museum for several years.

In 1958 when NASA was formed, my Naval Research Laboratory team became the core of the Goddard Space Flight Center. Over the next fifteen years at Goddard, I invented, designed, built, tested, and analyzed data from a series of electro-optical sensors that measured the orientation of spinning satellites and moon probes relative to the sun and earth. Based on this work, I was awarded a scholarship that enabled me to pursue what eventually became my lifelong scientific goal: *to understand how the brain works.*

In pursuit of this goal, I graduated from the Ohio State University with a masters degree in radio astronomy and control theory, attended the University of Southern California for background in neuroscience, obtained a PhD from University of Maryland in electrical engineering and robotics, and attended Johns Hopkins and the University of Maryland Medical School to study biophysics and neuroanatomy. I also worked at the National Institute of Health to learn how to monitor the activity in single neurons in the brain that respond to visual and tactile stimuli.

As an engineer, it is my conviction that you don't really understand something until you can build it. From the beginning, it has been my conviction that the principal function of the brain is to generate and control behavior. So I have pursued my background in control theory and robotics as an experimental tool for understanding the workings of the brain. I started at the bottom, in the spinal cord, which is the lowest level in the control hierarchy of the brain. The circuits in the spinal cord are basically servo control computers that translate commands for force and velocity into motion of the limbs and muscles.

The next step up the brain's control hierarchy is the mid-brain, where feedback from the spinal cord is combined with inertial signals from the inner ear and visual input from the eyes to generate precise, coordinated motion in all vertebrates from fish, to birds, to humans. The cerebellum enables a bird to fly at high speed through the branches of a tree without collision, and to perch lightly on a twig. It enables a squirrel to jump from branch to branch. It enables a batter to hit a pitch, and an outfielder to compute where the ball will come down. It enables a quarterback to throw a touchdown pass, and a wide receiver to catch a football with one hand while touching his toes to the ground.

In a 1967 book, *The Cerebellum as a Neuronal Machine,* Eccles, Ito, and Santagothai published data about the cerebellum that was sufficiently quantitative and anatomically accurate for me and David Marr to independently formulate theories of cerebellar function that have been combined to become a classic used by researchers studying the cerebellum to this day.

I used this theory of cerebellar function as a part of my PhD thesis to control a robot arm that could be commanded to perform and learn various coordinated motions. From this work, I patented a neural network called the Cerebellar Model Articulation Controller (CMAC) that was named by Industrial Research Magazine as one of the most significant inventions of the year 1976. CMAC has been used in many adaptive control systems and is still used by neural net researchers for building adaptive controllers.

CMAC has the ability to rapidly learn to compute a wide variety of control functions, and can be stacked in layers to plan and execute complex behaviors. This observation led to the development of a family of intelligent control systems for robots and automated manufacturing systems. In 1973, I left NASA and joined the National Bureau of Standards (NBS) where I teamed with Dr. Anthony Barbera to develop a family of intelligent real-time control systems (RCS). RCS was first used as the basic architecture for robots and machine tools in an Automated Manufacturing Research Facility at NBS. This captured the attention of Dr. Tony Tether, who was then a program manager at the Defense Advanced Research Projects Agency (DARPA) and later became DARPA Director. He suggested that we use RCS for controlling a group of combat aircraft. Unfortunately, this was too revolutionary at that time even for DARPA, so our project was redirected to use RCS to control multiple autonomous underwater vehicles. NASA took notice and contracted us to design an RCS controller for the Space Station Telerobotic Servicer. The US Bureau of Mines contracted us to design standards for automated underground coal mine systems. Over the years, RCS controllers have been developed for the US Postal Service to control general mail facilities and automated stamp distribution systems. Commercial RCS controllers have been developed for machine tools used by General Motors and Boeing, and for a variety of commercial water jet and flame cutting machines.

Along the way, I became interested in how to build really large robots, and invented a series of RoboCranes© that were cited by Construction Equipment and Popular Science magazines as among the one-hundred top products, technologies, and scientific achievements of 1992. In recognition of my contributions to manufacturing technology, I was named a "Hero of US Manufacturing" by Fortune Magazine in 1997.

During the 1990s, Dr. Barbera and I developed controllers for a series of Army Ground Vehicles culminating in the 4D/RCS reference model architecture for Intelligent Vehicle Systems that was adopted by the Army Future Combat System (FCS) for the Autonomous Navigation System to be used on all large FCS vehicles, manned and unmanned.

Over the years, I have authored more than 180 scientific papers, journal articles, book chapters, official government studies, and popular press articles on intelligent systems, robotics, and economic implications. I recently published several scientific papers on computation and representation in the brain that explore the possibility of reverse engineering the human brain. I have lectured extensively throughout the world on robotics and intelligent systems, am a member of the editorial board of the Wiley Series on Intelligent Systems, and serve on the editorial boards of six prominent journals related to intelligent systems and robotics.

Since the beginning with my research for Peoples' Capitalism in 1976, I have been deeply concerned about the economic, social, and political impact of the rapidly approaching advent of super automation. In particular, I have focused on the implications for jobs and income for the average worker. In 1976, this was largely a theoretical problem. In 2011, it has become a very current and practical issue. Modern industry simply does not need millions of new workers while the population of available workers is multiplying and the need for farm workers is declining around the globe.

In the 1976 book, I presented my vision of how the benefits of technological developments in advanced automation could become the means for liberating humankind from poverty and oppression. In this current book, *A Path to a Better World*, I have expanded this vision into a quantitative economic model for solving the current problems of slow economic growth and rising debt, and have suggested a foundation for a future prosperity based on widespread ownership of capital assets.

I was born to and raised by my parents, George and Lucy Albus, in Louisville, Kentucky. I received a BS in physics from Wheaton College (Illinois) in 1957, an MS in radio astronomy from Ohio State University in 1958, and a PhD in electrical engineering from the University of Maryland in 1972. I currently live with my wife, Cheryl, in Kensington, MD.

A brief third-party summary of my work is available at: http://en.wikipedia .org/wiki/James_S._Albus. A complete CV is available at: http://www.james-albus.org/docs/CV_10_29_10.pdf.

Outline

Chapter 1 offers a prelude to prosperity. Chapter 2 presents a vision of what a New World could be like, and contrasts that vision with the reality of the current situation. Chapter 3 discusses the mechanisms of free market capitalism that have been effective in generating economic growth, and analyzes flaws that permit poverty to exist alongside plenty. Chapter 4 suggests modifications to the current version of capitalism that would stimulate rapid economic growth and permit everyone to experience the benefits of owning a growing portfolio of capital assets. Chapter 5 reviews the technology that has brought us to the modern era, and previews future advances that could support prosperity for all. Chapter 6 proposes a new source of clean energy that would be environmentally friendly, carbon neutral, and virtually inexhaustible. Chapter 7 reviews the implications of intelligent systems technology for national security. Chapter 8 summarizes the effect of broadening access to ownership of capital assets on the general welfare in terms of personal freedom, economic security, domestic tranquility, and the pursuit of happiness.

ACKNOWLEDGMENTS

I want to thank my editorial advisor George Nedeff at iUniverse for his sage advice regarding the style and substance of this book. His suggestions have made the manuscript much more readable and tightly written than otherwise would have been possible. In addition, the working editorial staff have provided a wealth of positive suggestions regarding the organization of the material.

CONTENTS

CHAPTER 1 ⅠⅠⅠⅠⅠⅠⅠⅠⅠⅠⅠⅠⅠⅠⅠⅠⅠⅠⅠⅠⅠⅠⅠⅠⅠⅠⅠⅠⅠ■➤
Prelude to Abundance

INTELLIGENT SYSTEMS TECHNOLOGY IS DEVELOPING rapidly, and the rate of development is increasing exponentially. The next generation of advanced automation, particularly human-level artificial intelligence, raises serious social and economic questions. Truly intelligent machines will have the potential to eliminate poverty and usher in a new age of prosperity, opportunity, and economic justice. But they also have the potential to throw people out of work and widen the gap between rich and poor. A way to assure that the result is positive is to broaden access to credit for investment so that everyone can play the capitalist game with income from ownership of capital assets. This book is a plan to make that happen.

Our modern civilization is poised on the cusp of an information technology revolution that will at least equal, if not far exceed, the Industrial Revolution in its impact on humankind. The question is: Will the effect be for good or evil? What will the world be like when most goods and services are produced by intelligent machines? On the one hand, there is a possibility that exponential productivity growth could produce rapid economic growth, leading to an age of unprecedented prosperity. On the other hand, there are questions such as: Who will own these machines? How will displaced workers get an income? What is the potential for rising unemployment, poverty, and civil unrest?

From the beginning of human existence, mankind has lived under an ancient biblical curse: "By the sweat of thy face shalt thou eat bread, till thou return unto the ground."[1] In all the thousands of centuries prior to the

---

1   Genesis 3:19 (KJV).

Industrial Revolution, most of the human race lived near the threshold of survival. Virtually all economic wealth was created by the sweat and muscle power of humans and domestic animals. The substitution of mechanical energy for muscle power during the industrial revolution partially lifted the ancient curse. A little more than two centuries after the introduction of steam power into the manufacturing process, a large percentage of the population of the world lives in a manner that far surpasses the wildest utopian fantasies of former generations.

The technology of the Industrial Revolution, combined with the emergence of capitalism, has produced a dramatic increase in the rate of economic growth. This led to a degree of prosperity never before experienced during the entire history of the human species.[2] Indoor plumbing, cotton sheets and underwear, flush toilets, clear glass windows, central heating and air conditioning, automobiles, the telephone, radio, and TV are commonplace. Machines powered by electricity and the internal combustion engine largely replaced muscle power in agriculture, manufacturing, construction, mining, and transportation. An exponential increase in scientific and technological knowledge resulted in productivity improvements that have enabled the production of goods and services at a rate far in excess of any historical precedent.

The machines of the Industrial Revolution required human workers to control them, and the processes of commerce required humans to manage them. By virtue of their indispensability, workers have often been able to demand fair compensation for their labor through collective bargaining.[3] During the twentieth century, this led to the emergence of a prosperous middle class.

However, this is beginning to change. The information technology revolution is based on the substitution of computers for human brains in the control of machines and industrial processes. The application of information technology to the control of industrial processes and business management will bring into being a new generation of machines—intelligent machines that can create wealth largely unassisted by human beings.

There is good reason to believe that the information revolution will change the history of the world more dramatically than the Industrial Revolution. Factories and industries that can operate mostly unattended by human workers are already technically feasible and are becoming economically practical. Many of these factories even reproduce their own essential components. Machine tools are used to make machine tools. Computers are indispensable

---

2   Bronowski, Jacob. *"The Ascent of Man"*. Boston: Little, Brown & Co., 1973.
3   Ashenfelter, Orley C. and Richard Layard, eds. *Handbook of Labor Economics.* Amsterdam: Elsevier Science B.V., 1986.

to the design and manufacture of computers. Factories produce the structural materials and tools that are used to build new factories. Modern manufacturing machines and processes are able to construct extremely complex parts directly from computer data files. This suggests that manufactured goods may eventually become as inexpensive and unlimited by complexity as the products of biological mechanisms in living organisms. The potential long-run effects of this are profound and unprecedented.

The present economic system is not structured to deal with the implications of the coming generations of advanced automation. Classical economic theory is based on the labor theory of value. For the vast majority of the population, jobs are the primary source of income. The owners of the means of production represent only a tiny fraction of the population. This socio-economic structure is not well suited for a world where intelligent machines will perform most of the work necessary for the production of goods and services without human intervention. As machine intelligence grows, capital will replace labor as the principal factor in the production equation. The percentage of the population needed for producing all the goods and services that can be sold in the market will decline, unemployment will grow, and wages and salaries will experience steady downward pressures. Unless there emerges new mechanisms by which wealth can be distributed to average people, economic growth will slow, poverty will increase, and a few very rich owners of capital will grow fabulously wealthy.

The problem is that there presently exists no means by which average people can fully benefit from the unprecedented productive potential of truly intelligent machines—quite to the contrary. Under the present economic system, the widespread deployment of automated production systems will enable companies to lay off workers, reduce labor costs, minimize pension funds and increase profits in ways that threaten jobs and undermine the financial security of virtually every American family. In general, investors do not fund corporations to provide employment for workers. They are established to generate profits for owners.

There are two possible futures depending on how ownership of the means of production is distributed among the population.

One is to continue the current economic system wherein a small group of capitalists own most of the capital assets, and businesses increase their profitability by getting rid of workers. In this future, the gap between the rich and poor will grow, high unemployment will become the norm, wages will decline, and the number of people living in poverty will grow. Economic growth will slow as the number of jobs falls and consumer demand slows.

A second possibility is to adopt a new economic paradigm that takes advantage of the potential of technology development, and broadens the ownership of capital stock so that everyone can benefit from increasing productivity and profitability generated by advanced automation. This second possibility could enable the elimination of poverty and the creation of an era of prosperity and financial security for all humankind. It could provide a steady rise in consumer demand and generate rapid economic growth that would solve most of the problems of unemployment and rising debt that currently plague the economies of Europe, Japan, and the United States. It could reverse the decline of the middle class, and provide financial security based on ownership of capital assets to every individual.

The great challenge of the coming information technology revolution will be the development of an economic system wherein prosperity can be achieved and personal freedom can be preserved in a world where most wealth is created by intelligent machines. This book offers a plan by which this could be accomplished.

Specifically two new institutions are proposed:

1. **A Personal Investment Program (PIP)** is proposed to finance capital investment for increasing productivity in the production of goods and services. The PIP would authorize the Federal Reserve to issue credit to local banks for average citizens to invest in approved investment funds. These investments would spur economic growth and enable private industry to modernize their plants and machinery. Profits from these investments would be used to repay the loans and return dividends to the individual citizen investors. By this means, the average citizen would receive income from the industrial sector of the economy quite independently of employment in factories and offices. Every citizen would have the opportunity to become a capitalist in the sense of deriving a substantial percentage of his or her income from dividends earned on invested capital.

2. **A Personal Savings Program (PSP)** would be instituted in parallel with the PIP in order to provide sufficient savings to offset the inflationary effects of PIP investment spending. PSP savings would prevent short-term demand-pull inflation. The PSP would withhold income from consumers by mandatory payroll deductions and convert it into five-year personal savings bonds at market interest rates. Deductions would be graduated according to income (low-income persons would have little withheld, high-income more) and would be adjusted monthly according to a formula based on the best available indicators for inflation. The PSP would allow high rates of investment to

generate rapid economic growth and low unemployment while preventing excess demand from forcing prices upward.

I will argue in the following pages that, if implemented, these proposals would stimulate rapid economic growth and reduce unemployment in the short term, and in the long term (i.e., three or four decades) would lead to:

1. A society where every adult citizen would derive a significant fraction of his or her income from invested capital.

2. A society where ownership of the means of production would be distributed widely enough so that every citizen would be financially independent.

3. A society where many people would continue to work for supplemental income or the personal satisfaction that comes from contributing to society and being successful in the work place, but no one would be forced to work out of economic necessity.

4. A society where a diversity of lifestyles would flourish, opportunities for entrepreneurship would abound, and rewards for achievement would be high.

5. A society in which economic growth would be rapid, unemployment would be low, prices would be stable, and prosperity would be widespread.

In short, this book is a plan whereby Peoples' Capitalism[4] could be achieved in the United States before the middle of the twenty-first century without any significant changes in our constitutional form of government. In fact, far from altering any of the fundamental principles upon which this country was founded, this plan would revitalize the free-enterprise system and strengthen democracy. In the process, it would mobilize the full creative resources of our scientific and engineering capabilities to solve our most pressing human problems. The proposed economic system might best be described as a blueprint for Jeffersonian Democracy in a modern technological society.

---

4   Albus, James S. *Peoples' Capitalism: The Economics of the Robot Revolution.* Kensington, MD: New World Books, 1976. Available online at http://www. PeoplesCapitalism.org.

CHAPTER 2 IIIIIIIIIIIIIIIIIIIIIIIIIIIII➡
The Vision and the Reality

A T THE END OF THE first decade of the third millennium AD, the great economic and political debates center around economic philosophy and political ideology expressed in terms of left vs. right, liberal vs. conservative, Democrat vs. Republican, capitalism vs. socialism, government regulation vs. free markets. What is the proper role of government vs. the private sector? How big should the government be? How many taxes should the government collect, and from whom? Does the government have a responsibility for enforcing social and economic justice, as well as civil and criminal justice? How much should the government regulate the financial sector, the labor market, the consumer market place, the quality of the environment? And how should the economic problems of slow economic growth and rising debt be addressed?

Often lost in the heat of these arguments are more fundamental questions such as: What is the ultimate goal of the economic system? What kind of world do we really want to live in? What is the best way for wealth to be generated and income distributed? Given the state of scientific knowledge and technological progress, how fast could the economy grow? What is possible in terms of economic prosperity, environmental preservation, individual liberty, financial opportunity, and personal security? In other words, what is our concept of a good society?

I begin with my vision of what an ideal world might be like.

The Vision

A world without poverty

Imagine a world without poverty, hunger, or homelessness where everyone has a decent place to live, with plenty to eat. Imagine a world where everyone has access to a good education and adequate health care. Imagine a world where everyone lives in a safe, clean community without fear of crime or violence. Imagine a world where every human being can look forward to a secure retirement. Imagine a world where every disabled and elderly person can afford assisted living and dignified elder care.

A world of opportunity and prosperity

Imagine a world where opportunities abound to become rich, but no one is poor. Imagine a world where there is no ceiling at the top for how wealthy a person can become, but there is a floor at the bottom to prevent anyone from sinking into poverty. Imagine a world where jobs are plentiful in a wide variety of interesting fields, unemployment is low, and workers are well paid. Imagine a world where every individual has many opportunities to succeed and prosper.

A world of economic justice

Imagine a world where all are financially secure and many are wealthy. Imagine a world where economic growth is rapid and median income consistently grows at the same rate as the overall economy. Imagine a world where everyone benefits from productivity improvements because everyone owns a share of the means of production. Imagine a world where every individual owns capital assets that pay dividends in an amount sufficient to support a decent standard of living.

A world without pollution

Imagine a world where rapid economic growth is environmentally friendly: a world where the air is clean, the rivers, streams, and lakes are pure, the environment is preserved, wildlife is protected, and wilderness is preserved for future generations. Imagine a world where clean, renewable energy is inexpensive and effectively unlimited in supply. Imagine a world where society is both willing and able to afford the cost of a clean environment: a world where the energy to light, heat, and air condition homes, drive cars, run transportation systems, brighten cities, and power industrial plants is derived entirely from renewable sources that are carbon neutral, generate no significant air or water pollution, and create no risk of radioactive contamination.

A world without violence

Imagine a world where nations live in peace and prosperity without terrorism or war. Imagine a world where prosperous peoples rid themselves of dictators and tyrants. Imagine a world where people are too well educated and too financially secure to be seduced by bigots or fearmongers, a world where masses of poor and ignorant people no longer exist to be exploited by political radicals or religious zealots. Imagine a world where historical grievances have been relegated to the past, and tribal grudges have been replaced by individual hopes and dreams of opportunity, prosperity, and longevity for themselves and their children and grandchildren.

Is this vision only a dream?

Is poverty inevitable? Is hunger and homelessness unavoidable? Is inadequate medical care necessary? Is a world of peace and prosperity beyond the realm of possibility? Is modern technology unable to provide both economic prosperity and a clean environment? Are crime, violence, tyranny, and war predestined by some immutable law of human nature? Is there something in the human psyche that makes different tribes with different religions or conflicting ideologies incapable of living peaceably together without violence and bloodshed?

Many would answer, "Sadly, yes. Your vision is only a dream. We are entering an era of austerity, an age of declining expectations. The promise of rapid economic growth has proven illusionary. Poverty, crime, injustice, violence, and war are the way it has always been, and always will be. That is human nature. World peace and prosperity are the stuff of heavenly visions, holiday wishes, and 'Miss America' speeches."

Most people believe that poverty will always exist, along with crime, injustice, violence and war—and for good reason. The world has never experienced a period without violence and poverty. Nature itself is violent and survival in the natural world is a constant life-or-death struggle. Since the appearance of humans on this planet there has never been a time when most people were not poor. There has always been a war somewhere. Usually, many wars are in progress simultaneously. Most are tribal and civil wars over land, water, property, or religion; but history is also filled with wars between kingdoms, nations, and empires over power, dominion, and nationalistic pride.

Certainly there are historical grounds for pessimism. And this pessimism is deeply engrained in religious tradition. The Christian doctrine of original sin teaches that greed, envy, and injustice are fundamentally rooted in human nature. The Buddha teaches people to cope with adversity by repressing desires. Jehovah and Allah are tribal Gods that are portrayed in their respective holy books as caring more about their follower's religious devotion than their economic well-being. In every religion, there are close connections between the political and religious elites, and while there are many injunctions toward charity to alleviate suffering of the poor, none offer practical political plans for the eradication of poverty, at least not in this life. The promise of eternal bliss in cities with streets of gold is reserved for the afterlife as a reward for faith and devotion. The reality is that poverty has always existed. The vast majority of the world's population has always been poor, and most people believe that it always will be that way.

The Reality

There are more than 6.6 billion humans alive today. More than 40 percent of these are desperately poor. 2,560,000,000 human beings live on less than $2 per day.[5] That is more than eight times the total population of the United States. One in every seven humans, and one out every four children, live on less than $1 per day. Two hundred fifty million children live on the streets.[6]

In many regions, people have no clean drinking water. Many live without public sanitation. Hunger is a persistent problem. The United Nations estimates that about twenty-five thousand people die every day because of hunger or hunger-related causes. Around the world, millions of people live in refugee camps, displaced by civil wars, floods, earthquakes, tsunamis, and storms. Around and within the great cities of the world, there are slums where millions of people sleep on streets or rooftops, or live crowded together in shacks built of sheet metal, wood, and plastic retrieved from the local dump—without public water or sewer services, paved streets, or police protection. Many live as squatters on unoccupied land around industrial plants, garbage dumps, or hillsides that are unstable and unsafe for construction. China and India have the largest number of slum dwellers with 193 million and 158 million people, respectively. In many Third World countries, more than 80 percent of the

5   World Bank. *"World Development Indicators."* http://econ.worldbank.org/WBSITE/ EXTERNAL/EXTDEC/EXTRESEARCH/0,,contentMDK:22452035~pagePK:6 4165401~piPK:64165026~theSitePK:469382~isCURL:Y,00.html (accessed May 11, 2011).

6   Singer, P.W. *Wired for War: The Robotics Revolution and Conflict in the 21ˢᵗ Century.* New York: Penguin Press, New York, 2009, p. 285.

urban population lives as slum dwellers. UN researchers estimated that more than 1 billion people were living in slums in 2005.[7]

Even in America, 39 million people live in poverty.[8] On any single night in the United States, an estimated six hundred thousand people sleep on heating grates and park benches, in doorways, under bridges, and in shelters for the homeless. Many of the urban poor live on the streets, live in cars, or squat on unoccupied land in tents and shacks. During the harshest weather, they either migrate south, retreat to public shelters, or freeze to death.

Disparity in wealth and income

Throughout history, extreme wealth has existed alongside poverty. Although the vast majority of the population has forever been poor, there has always been a tiny minority that has managed to live very well indeed. From biblical times and before, wealth and power have been concentrated in the hands of a few while the large bulk of humanity has lived on the edge of survival. As civilizations grew and the range of control of kingdoms and empires expanded, the concentration of wealth and power grew, and opulence became more and more ostentatious. In ancient times the rich were pharaohs, sultans, princes, and emperors. In medieval times they were kings, czars, princes, and popes. During the Industrial Revolution they were industrial tycoons and robber barons. In modern times they are investment bankers, hedge fund managers, and corporate CEOs.

Today, a tiny percentage of the population receives incomes in excess of $100,000 per day. A few are superstar athletes or entertainers. Some are entrepreneurs. Most are corporate CEOs, Wall Street bankers, and hedge fund managers. In 2009 during the worst recession since 1929, twenty-five hedge fund managers had a combined income of $25 billion. Five had incomes in excess of $1 billion. The top earner had income of $4 billion.[9] Assuming he worked a forty-hour week, with two weeks off for vacation, that comes to $2 million per hour, or $16 million per day.

Something is wrong with this picture. This degree of disparity suggests that the current economic system is far from optimal in terms of the utilitarian

---

7   UN-HABITAT, *City Mayors Report,* February 2004: http://www.citymayors.com/report/slums.html.

8   DeNavas-Walt, Carmen, Bernadette D. Proctor, and Jessica C. Smith, US Census Bureau, Current Populations Reports P60-236, *Income, Poverty, and Health Insurance Coverage in the United States: 2008,* Washington, DC: US Government Printing Office, 2009. http://www.census.gov/hhes/www/poverty/poverty08/pov08hi.html.

9   Taub, Stephen, "Rich List," *AR Magazine,* April 1, 2010.

ideal of providing the greatest benefit for the greatest number of people.[10] Since 1980, top incomes in the US have exploded, while those of average workers have been flat, or have actually declined. As shown in figure 2.1, since 1979, income for the top 1 percent of the population has risen 281 percent whereas the income of the bottom 20 percent has risen only 16 percent.

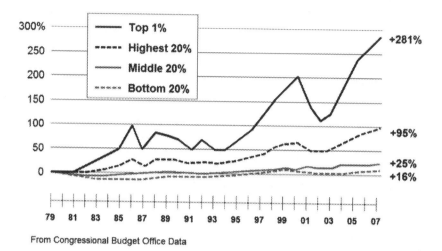

From Congressional Budget Office Data

Figure 2.1. Growth in income since 1979 for
four segments of the population.

America has returned to income disparity that is comparable to the 1920s. The top 1 percent of the households in the US receives 23.5 percent of the income, the highest percentage since 1928.[11]

Social and economic consequences

The problem with large disparity in wealth and income is that it breeds resentment. Poverty is a sign of lesser status, a lower rung on the social ladder. Wealth is a sign of greater status. In any social situation there are winners and losers. The winners accumulate wealth and power. The losers are forced to live in the most undesirable places and do the hardest and most dangerous physical work.

There are several ways that society has developed to keep resentment from

---

10   R. Harrison, "Jeremy Bentham," in Honderich, T., ed., *The Oxford Companion to Philosophy*, Oxford: Oxford University Press, 1995, pp. 85–88.

11   Piketty, Thomas and Emmanuel Saez, "Income Inequality in the United States: 1913-1998," *Quarterly Journal of Economics*, 118, no. 1 (2003). 1-41. Their most recent data set is available at http://elsa.berkeley.edu/~saez/TabFig2007.xls.

boiling over into civil disorder. One is raw political power. In the days of pharaohs and emperors, the rulers often claimed direct descent from the gods, and in some cases, declared themselves to be gods. During the era of kings and czars, extreme disparity in wealth was justified by proclamations of divine rights. Usually these claims were supported by the religious establishment, which was handsomely rewarded for its acquiescence and support. Armies were used both for warring with other countries and for suppressing internal dissent.

Since the Industrial Revolution, great disparity in wealth has been justified on economic grounds.[12] Benefits to the average citizens have been confined mostly to improved quantity, lower prices, and higher quality of products. The owners of the means of production have reaped the monetary benefits which they have been reluctant to share with the workers. Wages are paid on the basis of market supply and demand for labor. Henry Hazlitt, in his best-selling economics book *Economics in One Lesson,* argues that

> "while labor organizations may limit the supply of workers which results in higher demand and greater incomes for members. In effect, some workers end up with higher than market wages at the expense of their newly unemployed co-workers. This creates an economic inequality in itself. Members may also receive higher wages through collective bargaining, political influence, or corruption."[13]

During the nineteenth and twentieth centuries, the very rich were industrial tycoons and financiers. Today, the top of the wealth pyramid is occupied by corporate CEOs and Wall Street money managers. In modern times, displays of great wealth have become less ostentatious and largely hidden from the poor. The rich live in walled estates mostly unseen by the masses. They anchor their yachts in exclusive ports where they can socialize with their own kind.

Occasionally, liberal political reforms have forced the captains of industry to share some portion of the wealth. During the last century, union movements, labor shortages, alternatives for workers provided by an expanding frontier, and democratic forces within society compelled management to share some of the profits with the workers. This gave rise to a middle class that provided political stability. In the democratic industrialized world, the promise of opportunity and upward mobility has created a situation where the average

---

12   Rand, Ayn. *Atlas Shrugged*. New York: Penguin, 1957.
13   Hazlitt, Henry. *Economics in One Lesson,* New York: Three Rivers Press, 1988.

person believes that with hard work and ambition, they too can become rich, or at least move up the ladder of prosperity.

However, in the slums of the world, there is no such optimism. More than a billion people live in slums that surround and infiltrate the big cities of the world.[14] Unemployment is rampant. Most of the population is illiterate or semiliterate. Only the most menial jobs are available. Hope for climbing up the ladder of prosperity is dim at best. These slums are growing rapidly as surplus workers from farms and villages migrate to urban areas even as cities cease to be job machines.[15] In 1950, there were eighty-six cities in the world with a population of more than one million; in 2006 there were four hundred, and by 2015 there will be at least five hundred.[16] In 2006, there were eighteen cities with more than 10 million residents. Slum residents made up an astonishing 78 percent of this population.[17] In his book, *Planet of Slums*, Mike Davis estimates that there are more than two hundred thousand slums in the world, ranging in population from a few hundred to more than a million persons. Five cities in South Asia (Karachi, Mumbai, Delhi, Kolkata, and Dhaka) alone contain about fifteen thousand distinct slum communities whose population exceeds 20 million.[18]

Slums are breeding grounds for disease and violence. They are fertile ground for insurgencies and terrorism. Urban zones are overflowing with angry, disenfranchised youth that are accustomed to violence, easily radicalized. These provide a large pool of unemployed and uneducated young men and women who can be easily recruited by criminal gangs, drug lords, death squads, revolutionary cells, and terrorist groups. In Bogota, Columbia, businessmen vandalized by street urchins form links to right-wing death squads, and the bodies of murdered children are dumped at the edge of town.[19] On the outskirts of Santiago de Cali, the "garbage mountain" slum of Navarro is where hungry women and children pick through garbage, while youthful gunmen are hired by local militia to extort money from the poor and kill youths from other militia in battles over territory.[20]

A World Bank report in 2000 warned that urban poverty will become

---

14   UN Habitat, 2003, "Slums of the World: The Face of Urban Poverty in the New Millennium," working paper, Nairobi, annex 3.

15   Davis, M. *Planet of Slums*. New York: Verso, 2006, p.15.

16   Ibid, p. 1.

17   Ibid, p. 23.

18   Ibid, p. 26.

19   Project Counseling Services, "Deteriorating Bogota: Displacement and War in Urban Centers," *Colombia Regional Report: Bogota,* Bogota, Columbia: December 2002, pp. 3–4.

20   Taussig, Michael. *Law in a Lawless Land: Diary of a Limpreza in Colombia.* Chicago: The University of Chicago Press, 2003, pp. 114–15.

the "most significant and politically explosive problem of the 21st century."[21] US military experts predict that feral, failed cities of the world will be the battle space of the twenty-first century. Military planners expect urban slums to be where the wars of the future are going to take place. That is where the enemy will live. [22]

In Muslim slums, religious charities and extremist clerics provide what little education is available. Young men are steeped in myths of holy prophets and martyrs, and their minds are filled with tales of religious battles between true believers and infidels. Their concept of history is colored by a litany of ancient and recent tribal battles and atrocities.

Yet even in the poorest slums, the youth are informed about the outside world. They watch television. They attend movies. They get information and reinforcement from others on the Internet. Principles of democracy are widely known, if not followed in practice. Today's losers are no longer ignorant of the gap between rich and poor. And the losers are more numerous and much better armed than ever before. They have easy access to guns. Many of them live in neighborhoods where even the police dare not go, especially at night. Many live where the police show up only to collect bribes.

The unemployed youth in today's exploding slums provide a vast boiling sea of willing recruits for political and religious extremists. Today's war lords, gangs, and terrorist groups are more sophisticated than ever before. They are better able to organize because of the Internet and cell phone technology. They are better armed. Automatic weapons are readily available. They have access to assault weapons, rocket-propelled grenades, and explosives for improvised bombs. And they are well financed. Radical charities collect money from sympathetic wealthy donors and purchase weapons for terrorist groups. The drug trade provides a voluminous source of income that is used to pay private armies, corrupt government officials, and bribe police. Antiaircraft missiles can be purchased on the black market. There are state governments that support terrorist groups and use them as proxy armies. It is only a matter of time before a terrorist organization gains access to tactical nuclear weapons. We can no longer assume that the poor will remain passive. It is imperative that we address poverty as a fundamental cause of terrorism, and therefore a national security issue, before it is too late.

The dangers of poverty are often dismissed on the grounds that people in

---

21   Shi, Anqing. "How Access to Urban Potable Water and Sewerage Connections Affects Child Mortality," Finance Development Research Group working paper, World Bank, January, 2000, p. 14.

22   Singer, P.W. *Wired for War*: The Robotics Revolution and Conflict in the 21st Century. New York: Penguin Press 2009, p. 289, quoting military expert Ralph Peters and urban theorist Mike Davis.

the deepest poverty tend to be submissive and too preoccupied with surviving to participate in protest movements or political organization. The very poor may opportunistically participate in looting or rioting as the occasion permits, but they seldom pose a serious threat to political stability. Civil unrest can usually be contained through police force, so long as the economic system retains at least some semblance of fairness and economic justice. Occasionally, riots break out triggered by some spark of outrage, such as a police shooting or videotaped beating, but ordinarily, most of the violent crimes are confined to gang wars within the slums themselves.

However, the leadership for radical movements typically does not come from the poorest of the poor. It is the better-off members of the tribe who become outraged by perceived injustice to their poor and oppressed brothers and sisters, and aspire to political power by leading revolutionary movements. It is the more affluent members of the community that have the education and economic resources to publish manifestos and organize resistance movements. It is people like Lenin, Mao, Castro, Che Guevara, and Osama Bin Laden that have the time and money to dedicate their lives to revolution and the formation of terrorist cells.

The traditional survival strategy for the desperately poor is to have many children that can be put to work early and grow up to support the family elders. Unfortunately, in the vast urban slums of the world, this strategy simply adds to the exploding population of unemployed youth that are susceptible to gang violence, indoctrination into terrorist organizations, and often unwilling conscription into private armies. Idealistic youth from the slums can sometimes even be recruited as suicide bombers if promises are made of paradise in the afterlife, and rewards are paid to their families in this life. Thus, extreme poverty provides foot soldiers for insurrection and terrorism.

Of course, there are many causes of conflict other than poverty. There are ethnic and religious hatreds and battles over territory. However, the underlying cause of most of these conflicts is poverty. People with nothing have nothing to lose. Prosperous people typically do not fight wars with other prosperous people over religious differences. If the well-to-do have disputes over property, they hire lawyers and fight their battles in court. The rich engage in competition in the marketplace or political arena. But they don't usually resort to physical violence because they have too much to lose. Rich countries may send armies to invade other countries, but usually only those that are relatively defenseless and unable to strike back in a meaningful way. In prosperous communities, political and religious conflicts are resolved without violence, if for no other reason than fear of retaliation. Tribal wars, such as are regular events in Africa, the Middle East, and Asia, are fundamentally fights between poor people over the basics of survival (i.e., land and water).

Reasons for Hope

Yet, while there certainly are reasons for pessimism, there are also reasons for hope. Prosperity can cure many ills and smooth over many grievances. We have the physical capacity and production technology to create a decent world for all humanity. The exponential advance of technology suggests that the possibility exists for productivity improvements that could support an era of prosperity that would bring peace for all humankind. The demonstrated capacity of free market capitalism to satisfy consumer demand offers hope that all the world's people could be supported in a comfortable and prosperous lifestyle. We have the technical and economic tools to support a much more rapid rate of economic growth. We have the capacity to eliminate poverty and create a world of prosperity in an environmentally friendly way. We could do much better.

This is not the best of all possible worlds. The current capitalist economic system contains a number of contradictions and counterproductive features that keep it from performing as well as it could. Throughout the remainder of this book, I will examine some of these features, and suggest some policy changes that could be implemented to fix them. In short, I will lay out a plan by which the vision enunciated at the beginning of this chapter could be achieved.

CHAPTER 3 ||||||||||||||||||||||||||||||||➡
Free Market Capitalism

C APITALISM IS AN ECONOMIC SYSTEM characterized by private or corporate ownership of capital goods by investments that are determined by private decision rather than state control, and by prices, production, and the distribution of goods that are determined mainly in a free market.[23] In a capitalist economy, individuals and firms have the right to earn income and to purchase labor for wages. Capitalism is based on a free market with open competition. Regulation is achieved mainly through the operation of market forces where prices and profits dictate where and how resources are used and allocated.

The modern capitalistic economic system has clearly demonstrated the ability to produce goods and services more efficiently, in greater quantity, and with better quality than any other type of economic system in history. Over the past two centuries, free market capitalism has created a vast explosion of wealth and prosperity that is beyond the wildest dreams of optimistic visionaries two hundred years ago. Today we take it for granted. We have luxurious homes, fine furnishings, expensive cars, high-definition television, cell phones, personal computers, the Internet, and the ability to travel anywhere in the world in a matter of hours. How was this accomplished? What is the secret?

First of all, there is the principle of private ownership. Private ownership creates incentive for individuals and corporations to acquire productive capital, because the profits accrue to the owners. It creates incentives for innovation, because more productive machines and more efficient methods produce more

---

23   Webster's dictionary definition.

and better output, and thereby generate more profits. It creates incentives for reinvesting profits back into the corporation to improve productivity, increase capacity, and lower costs in pursuit of greater market share.

Secondly, there is the market. Competition for customers in the marketplace creates incentives for high quality goods, low prices, and good customer service. Consumers can shop for the best quality at the lowest price. Producers must compete for customers if they want to sell their goods and services. The result is that supply meets demand at the lowest price to the consumer. Competition between producers drives prices down. This puts pressure on producers to cut costs and improve efficiency. To survive in the marketplace, producers must innovate, so that they can improve their products and lower their costs. Those companies that do not innovate soon find themselves unable to compete in the market. Companies that sell obsolete products, sell inferior goods, or fail to minimize costs go out of business. Thus, the free market rewards innovation, punishes inefficiency, improves quality, and keeps prices low.

Third, there is the profit motive. Free market capitalism allocates resources efficiently. Companies strive for efficient use of time and materials. Those enterprises that are most efficient are profitable. They prosper and grow. Those that are less efficient fail or are unprofitable, and are quickly replaced by their more efficient competitors. Competition between producers drives companies to seek the most efficient and effective means of production, the cheapest materials, and the lowest-cost labor.

Fourth, there is the ability to respond quickly and effectively to market demand. Where there are customers with money to spend, capitalism will quickly be there to meet demand. The ability of capitalism to produce goods and services is constrained mainly by market demand. Almost all successful corporations either have or could quickly acquire the capacity to produce more products, if there were more buyers. The reason they don't is because it makes no sense to produce products that can't be sold. If customers are not willing or able to buy more products, companies must cut back on production.

To stimulate consumer demand, corporations rely on advertising. Over the years, advertising has become an essential element in modern capitalism. Advertising not only informs potential customers of new products and services, but actually creates demand for products that never existed before. This enables companies to introduce new products, expand demand for existing products, and attract more consumers.

In a capitalist economy, capital flows to wherever there is the potential to satisfy market demand for profit. Whenever market demand is growing, capitalism increases production to take advantage of new opportunities for profit. Whenever productive capacity is insufficient, companies invest in additional capacity to meet the growing demand. Shortages of consumer products are quickly overcome, and companies compete for their share of the market. In the world of free market capitalism, goods and services are plentiful and the customer is king.

It is commonly observed that the consumer spending is the largest component of the GDP. If consumers stop spending, the economy goes into recession. Consumer credit is a major factor in consumer demand. Banks and businesses issue credit cards and offer easy loans to induce consumers to buy more than they could afford if they had to pay cash. The government subsidizes consumer loans for purchasing homes.

Evidence for the success of capitalism can be observed in any shopping mall or supermarket in any capitalist country. Shelves are crammed with an abundance of goods that are attractively packaged to entice consumers to buy them. Automobile showrooms and dealer lots are filled with cars for sale. Businesses compete fiercely for customers. There are plenty of goods available, and manufacturers could easily increase production if there were greater consumer demand. The construction industry would gladly produce more homes, if there were more buyers. The farmers could grow more crops, if there were more shoppers at the supermarkets. In most advanced countries, the government pays farmers to not produce too much food so as not to drive prices down. The most common cause of business failure is not an inability to produce more goods or services, but an inability to attract customers that will buy their products at a profitable price.

History

Capitalism emerged from mercantilism which itself evolved from feudalism.[24] Under feudalism, society was organized around the manor, which was property granted to the local lord by the king in return for the lord's pledge to support the king in wars (which were frequent). The lords granted land to knights in return for their pledge to fight for the lord at his command. The serfs were bound to the land and did all the real work, and served as foot soldiers in the king's wars. The serfs agreed to serve the lord in return for protection from bandits and the right to work on the lord's land. The only real difference between serfs and slaves was that serfs could not be sold separately from the land to which they were bound. The feudal lords lived comfortably by the labor of the serfs, and gained wealth through political intrigue and military adventures.

The manor was a largely self-sufficient entity. Most of what was needed

---

24   Stephenson, Carl. *Medieval Feudalism*, New York: Cornell University Press, 1942.

was produced within the territory of the manor by the serfs. Nothing was produced for market, and very little was purchased. There was effectively no trade. The serfs had no incentive to increase production because the lord took all of the surplus for himself. Thus, productivity growth was imperceptible, and economic growth depended mainly on military conquest.

During the Renaissance, larger and stronger armies prevailed. Wealth and power became consolidated in larger city states, kingdoms, nations, and empires. With the increasing size and scope of political entities, industries emerged to provide goods for sale. Trade grew in magnitude and scope, and mercantilism emerged.

Mercantilism was a system where the military power of the state was employed to support economic growth.[25] Under mercantilism, the government of the king controlled the economy and used it to increase his power at the expense of rival nations. Armies were raised to seize territory from neighbors and establish colonies in underdeveloped parts of the world. State power, conquest, and acquisition of overseas colonies became the primary source of economic growth, and principal aim of economic policy.

Manufacturing was largely dedicated to production of weapons, but civilian production was also encouraged as a source of tax revenue for the government. Taxes were typically high and largely used to finance armies and navies, or to pay war debts from past military operations. Very little was reinvested in industry. If a state could not supply its own raw materials, it would send military forces to acquire colonies from which they could be extracted. Colonies constituted not only sources of supply for raw materials but also markets for finished products. Because it was not in the interests of the state to allow competition, colonies were prevented from engaging in manufacturing or trading with foreign powers.

Mercantilism became the dominant economic theory and policy in Europe from the sixteenth to the eighteenth centuries. This led to a virtually constant series of wars and shifting alliances between the European powers, and created resentment and conflict between the European powers and their colonies.[26] In the case of Britain and its North American colonies, this resentment led to the American Revolution.

The development of strong national states during the mercantilist era provided the trading relationships, monetary systems, and legal codes necessary for the rise of capitalism.[27] The constant need to raise armies and procure weapons led to the establishment of a banking system that was

---

25   Landreth, H. and D. C. Colander. *History of Economic Thought* (4th ed.), Boston: Houghton Mifflin, 2002.

26   Kennedy, P. *Rise and Fall of the Great Powers.* New York: Random House, 1987.

27   Ibid.

sufficiently independent and powerful to issue loans to governments to finance their wars. Thus, mercantilism provided the institutions and practices that supported the early development of capitalism.

During the eighteenth century, mercantilism was gradually replaced by laissez-faire capitalism as more and more privately owned corporations were chartered by the kings. These corporations began to invest profits back into the enterprise instead of spending them on palaces and cathedrals. This enabled capitalism to generate rapid economic growth. Because of their greater productivity and efficiency, privately owned corporations that could be taxed largely replaced state-controlled companies.

The rise of capitalism led to the emergence of entrepreneurship. It was enabled by the technology that flowed from the Industrial Revolution, and was aided and abetted by the Protestant Reformation, which sanctioned individuality, hard work, and frugality. The social customs of thrift and saving became emphasized as virtues because they made possible the accumulation of capital. With the end of mercantilism, capitalism became the dominant economic philosophy in the Western world, and private corporations became the primary source of economic growth.

Privately owned corporations designed to produce profits for the owners are the foundation of capitalism. Corporations may be owned by individuals or families, or may be publicly owned by stockholders who own shares of the corporation. For large corporations, stockholders are mostly banks, insurance companies, pension funds, wealthy individuals, and mutual funds. Mutual funds allow small investors to pool resources and spread risk by diversification over a large portfolio of individual stocks.

The Workers

In all economic systems, workers are needed to actually produce goods and services. In the ancient world, most of the work was performed by slaves. Under feudalism, most of the work was done by serfs. With the end of serfdom in Europe and slavery in the New World, most of the work is performed by free workers, who are paid for their labor and are free to quit, but have to fend for themselves for food and shelter.

Under capitalism, paid workers are needed to produce the goods and services that businesses produce for sale in the market. Farms need workers to plant and harvest crops, mend fences, tend to livestock, and maintain equipment. Factories need workers to design products, operate machines, maintain inventory, schedule production, assemble, test, package, and ship products. Construction companies need workers to operate equipment and build homes, office buildings, factories, roads, bridges, tunnels, and ports.

Transportation, wholesale, and retail companies need workers to distribute products and deliver goods to customers. Service industries need workers to provide police and fire protection, maintain utilities, clean buildings, maintain yards, and collect trash. Schools need teachers and administrators. Hospitals need doctors, nurses, technicians, and staff to care for the sick, elderly and infirm.

Under capitalism, there is a market for labor, just as there is a market for goods and services. And just as competition between companies drives down prices, competition between workers in the labor market drives down wages and salaries. The history of capitalism is filled with episodes where companies have exploited the labor market to keep wages low. On the other side, unions and professional societies are designed to reduce competition between workers, and force wages up.

But fundamentally, capitalism is an economic system fashioned by business owners for business owners. It is run by bankers, investors, managers, and executives for the purpose of making money for themselves and the stockholders. Successful corporations are those that generate the most profits. Workers are hired only when necessary to achieve profitability. Workers are paid only as much as is necessary to make the business operate efficiently. To the owners, wages and salaries are costs to be minimized whenever possible. This means that successful corporations hire only the minimum number of employees necessary to operate their businesses efficiently, and pay their workers as little as possible.

Of course, capitalism is not the only system that pays workers as little as possible. In feudal Europe, serfs were paid nothing and were compelled by law to work for their masters. In both the ancient and new worlds, slaves were paid nothing, and were bought and sold like livestock. Under mercantilism, labor costs were minimized and unions were forbidden by law. In the early days of capitalism, trade unions remained illegal. In Europe, workers were forced to work long hours, seven days a week for starvation wages. Children were used to pull carts of coal and ore through narrow tunnels in mines, or were employed in mills and shops where many were maimed or killed in accidents and fires. In the new world, wages were somewhat higher because the frontier provided an alternative to working in factories and mills. However, even in America, workers were paid as little as possible and often forced to work under unsafe and unhealthy conditions. Mine employees and railroad construction crews were often housed in company towns where the company owned all the houses and the only store was a company store. Rent and prices were high, and companies often offered loans at high interest rates to workers who struggled to make ends meet. This kept workers deeply in debt, and thus not free to quit or protest.

It was not until workers organized into unions to demand higher wages and better working conditions that some of the wealth generated by the capitalist economy began to trickle down to the workers. The history of the union movement is filled with conflict and confrontation. Early efforts by workers to organize into unions and strike were met with violence, both by the company management and occasionally by government troops.

It was not until the nineteenth century that laws were passed that gave workers a legal right to organize into unions to demand better wages and working conditions. In the early 1800s, workers in America made some short-lived efforts to organize into unions, but it was not until 1880 that the first successful union, the Knights of Labor, was formed. Six years later, the American Federation of Labor (AFL) was formed.[28] In 1891, Pope Leo XIII issued an encyclical entitled "Rights and Duties of Capital and Labor," which spoke against the atrocities workers faced and demanded that workers be granted certain rights and safety regulations.[29]

Most advanced industrial countries now have laws that protect worker rights to organize into unions and strike to force companies to negotiate fair compensation for their labor. But even today, corporations tend to fight union demands and move production facilities to places where union influence is limited and the cost of labor is minimal.

Communism

It was the exploitation of workers under mercantilism and early capitalism in England and Europe that motivated Karl Marx to develop his concept of communism as an alternative to capitalism. In *Das Kapital*[30] and other writings, Marx developed the economic theory that subsequently inspired communist revolutions in Russia, China, Cuba, Vietnam, Cambodia, and North Korea.

Communism is the antithesis of free market capitalism. Communism is an economic system in which the means of production and distribution are owned by the community, not by individuals. In a communist economy, prices are regulated by the community governors, goods and services are

---

28   Foner, Philip S. *"History of the Labor Movement in the United States: Vol 1." From Colonial Times to the Founding of the American Federation of Labor.* New York: International Publishers, 1978.

29   Pope Leo XIII, *Rerum Novarum, Encyclical on Rights and Duties of Capital and Labor.* 1891. http://www.vatican.va/holy_father/leo_xiii/encyclicals/documents/hf_l-xiii_enc_15051891_rerum-novarum_en.html. (accessed May 29, 2011).

30   Marx, Karl. *Das Kapital.* Moscow: Progress Publishers, 1887. Available online at http://www.marxists.org/archive/marx/works/1867-c1/. (accessed May 29, 2011).

subsidized, and businesses are operated not-for-profit. The community government decides where and how resources are allocated.

As opposed to capitalism, communism was, at least in theory, designed to benefit the poor. The origins of communism are much older than capitalism, and long predate Karl Marx. Jesus counseled a rich man who wished to inherit eternal life with the words: "sell all that you own and distribute unto the poor, and come follow me."[31] In the first Christian church, members sold their possessions and goods, and gave the proceeds to the Church to be distributed to those in need.[32] In succeeding centuries, monks and nuns in Christian monasteries have chosen to relinquish all earthly goods, work for the common good, and care for the poor. Thomas Moore's Utopia was about an imaginary communist community, which looked very attractive to poor people living in medieval England. In America, many utopian groups such as the Shakers have based their faith on communal Christian principles.

In theory, the communist principle of community ownership of capital assets creates an incentive for sharing. Marx wrote: "From each according to their abilities – To each according to their needs."[33] This notion appeals to the human sense of fairness and justice. Not surprisingly, it is popular among the poor. It still inspires revolutionary movements in many parts of the world today, including South America, the Philippines, the Mid-East, Southeast Asia, and Africa.

The problem in practice is that community ownership destroys incentives for individuals to work hard and be thrifty. Why work hard and save money if the fruits of your efforts are going to be diluted by being shared with everyone? Benefits to an individual from thrift and diligence are divided by the number of people in the community.

Similarly, the economic cost of laziness and wastefulness by an individual is diluted by the number of people in the community. Why work hard if you receive almost the same benefit from doing nothing? This tendency of human nature can only be countered by coercive means, or by religious fervor. It is not coincidental that the only truly successful and enduring communes have been strict religious orders.

In society at large, communist economies tend to be inefficient and lethargic. There are few incentives for risk-taking or productivity improvement. There are few personal benefits for individual achievements. In order to

---

31   Luke 18:18–2 (KJV).

32   Acts 2:44–45, Acts 4:34–37, and Acts 5:1–11 (KJV).

33   Marx, Karl. *"Critique of the Gotha Programme,"* in *Marx/Engels Selected Works, Volume Three*, pp. 13–30. Moscow: Progress Publishers,1970. Available online at http://www.marxists.org/archive/marx/works/1875/gotha/index.htm. (accessed May 29, 2011).

motivate economic activity, some form of authoritarian incentives must be imposed on the workers by the community embodied in the state. Thus, communist governments inevitably devolve into totalitarian states such as the Soviet Union under Lenin and Stalin, China under Mao, Cuba under Castro, North Korea under Kim Jung Ill, Vietnam under Ho Chi Min, or Cambodia under Pol Pot.

In communist countries, industries focus on providing jobs, rather than maximizing productivity. Everyone has a job, but the system is inefficient and productivity is low. Thus, economic growth is slow and there are few consumer goods to buy in the stores. This means that communist economies cannot rely on the market to set prices. Otherwise, shortages in supply would drive prices up and out of the range of the workers' incomes. Thus, communist economies rely on government price controls and rationing. Goods are cheap, but in short supply. People often must stand in long lines just to buy basic food items. The availability of middle-class goods like comfortable homes, cars, appliances, and leisure activities are unavailable, except to high-level party officials.

The lack of a market means that resources are not allocated efficiently. State-owned companies are not motivated by profits to make efficient use of time and materials. Their primary motivation is to provide jobs for workers regardless of efficiency. Thus, companies do not innovate and do not respond to consumer demand. Instead they continue to produce inferior products for which there is little demand in order to preserve employment for workers that have little incentive to work. As a result, there is very little value added per worker hour.

It is thus not surprising that communism is no longer a serious contender for economic growth in the major industrialized nations, or even in emerging economies such as India and China. Marxism still has appeal to the poor and disenfranchised around the world, particularly in countries where there exists an enormous gap between rich and poor. But the economic debate between capitalism and communism is over. Capitalism has won because it works. It produces economic wealth and rapid economic growth. Communism does not.

Defects in Capitalism

However, despite the triumph of capitalism over communism, there are several features of free market capitalism that one might classify as defects. Outperforming communism is not a sign of perfection. Producing better than catastrophic results is not a very high standard of excellence. That free market capitalism is the best system yet devised does not imply that it is the best system that could possibly be. It merely means that it is better than all the worse systems that have been tried before.

Hardly anyone is completely happy with the current economic system. In 2008, the Western capitalist countries experienced a near collapse of the financial system, and barely avoided a second Great Depression. Several European countries teetered on the brink of insolvency. Three years later, the United States remains stuck in a jobless recovery. In both the United States and Europe, GDP is growing slowly, but unemployment remains unacceptably high. Businesses are discovering that they can increase output without hiring workers—so they don't. Investors are reluctant to invest in new productive capacity because customer demand is low. Many workers have lost their jobs, and with them their purchasing power. Others are afraid they may lose their jobs so they have curbed their spending. Still others have maxed out their credit, and are cautious about taking on more debt. Economic growth is slow and unemployment is unacceptably high. Many people are losing their homes and sinking into poverty.

This suggests that there might be a few features of the current version of free market capitalism that could be improved.

Poverty

Perhaps the most glaring and unattractive feature of free market capitalism is the persistence of poverty. Despite the obvious fact that capitalism is capable of producing many more goods and services than it does, poverty is widespread throughout the world. For two hundred years, capitalism has demonstrated the ability to generate goods and services wherever there is consumer demand —and it clearly could produce more, if consumer demand were to increase. Capitalism's capacity for creating wealth is not constrained by limits on production, but by limits on customers in the market. Companies would gladly expand production if there were more customers to buy what could be produced.

Yet half of the people alive today remain desperately poor. Forty percent of the world's population lives on less than $2 per day. Even in the best of times in the most industrialized countries, poverty is a persistent problem. Why is this? The poor are in need of everything, from the basics of food, shelter, clothing, sanitation, clean water, police and fire protection, to more upscale consumer goods and services such as health care, education, appliances, cars, vacations, and retirement services. The poor of the world represent a huge potential market. If the poor had decent incomes, they would buy cars, clothes, homes, furniture, and appliances. They would pay taxes to obtain better utilities and improved public services. They would attend schools and universities. They would have fewer children and live in better neighborhoods. They would buy health insurance. They would travel and engage in leisure activities. But they do not, because they have no income.

If the poor had income, businesses would meet their needs. Corporations would expand production to supply them with products and services. Economic growth would be rapid. Business opportunities would explode and stockholders would reap huge profits. Why don't they? Why hasn't the world's industrial capacity simply been expanded to produce enough wealth for everyone to have enough to eat, a safe and decent place to live, good sanitation, adequate medical care, and a basic education? Why does the productive potential of modern technology remain underutilized? Why can't the productive capacity of free market capitalism meet the needs of all the world's people? The poor are potential consumers. Why doesn't capitalism expand output to meet their needs?

The obvious answer is the poor have no money. People with no disposable income generate no consumer demand. Capitalism responds to market demand, not to human need. Where there is no consumer demand, capitalism does not expand production. Adam Smith's "invisible hand" doesn't work its magic for people with no income.[34]

The truth is, capitalism was never designed to benefit the poor, or for that matter the working class. Capitalism is designed to benefit the owners of capital. To the capitalist system, the poor are invisible, except when they spoil the view by sleeping on park benches and heating grates. To the capitalist, the slums that surround all the world's biggest cities are an embarrassment, and an indictment. The poor are a denial of the claim that capitalism is the best of all possible economic systems. It is not surprising that many see capitalism as a system of domination that exploits the weak and ignores the poor.

Owners vs. workers

The tension between owners and workers is among the most fundamental structural features of capitalism. To maximize profits, companies must minimize costs. In most cases, wages and salaries make up a large portion of the cost of production. Thus, for any individual company, there are strong incentives to minimize labor costs.

However, to make any profits at all, companies must sell their products in the market. Without market demand, companies can't stay in business. The

---

34   The "invisible hand of the market" is the term economists use to describe the self-regulating nature of the market place. This is a metaphor first coined by the economist Adam Smith in *The Theory of Moral Sentiment.*

inability to sell products and services at a profit in the market is the primary cause of business failures.

Market demand is determined by how much money customers have and are willing to spend, and most customers are workers. Thus, if workers are paid too little, or if the number of jobs is too small, consumer demand in the market will fall. Wages and salaries paid to workers provide income that generates consumer demand in the market. Without jobs, workers cannot buy the goods and services that are produced, and without customers, companies cannot expand production.

The problem is that for any individual company, profits go up when labor costs go down. Corporations are designed to maximize income to the stockholders, not to create jobs or provide income for workers. However, for the economy as a whole, market demand depends on consumer income, which depends mostly on wages and salaries paid to workers. Thus, what is good for the individual company is bad for the economy as a whole.

This is a fundamental structural contradiction in the capitalist system. It may help explain why poverty persists despite capitalism's obvious capability to produce more. The poor don't create demand in the market because they have no money. They have no money because they don't have jobs. But they cannot get jobs because they are not needed as workers to produce the goods and services that they cannot buy because they don't have jobs. This is a vicious circle that is built into the current embodiment of capitalism.

Tendency toward monopoly

A third imperfection in capitalism is its tendency toward monopoly. Competition in the marketplace favors winners. Big companies have many advantages over small ones.

- They can employ economies of scale to exploit existing technology.
- They can engage in proprietary research to develop and patent improved technology.
- They have large cash flows and access to large amounts of credit from banks for investing in bigger and more efficient plants and equipment.
- They can buy potential competitors.
- They can outsource work to places where labor costs are minimal.

As corporations win in the market, they grow in wealth and power. The winners grow, the losers shrink and die. Eventually, unless the government intervenes, a single winner takes all, and monopoly is achieved. Once competition has been eliminated, the winner can raise prices to whatever the market will bear. Consumers then have no choice but either to pay the asking price, or go without the product. Without competition, the free market does not work to the benefit of the consumer.

Near the end of the nineteenth century, the big corporations and big financial institutions totally dominated the market. Railroads, oil companies, steel companies, mining companies, big banks, and trusts regularly conspired to eliminate competition and control the marketplace. The big corporations routinely used predatory business practices to eliminate competition. They would buy up competitors, and if the competitors refused to sell, they would drive them out of business by temporarily cutting prices so low that there were no profits to be made. The number of owners became fewer and fewer, and those that rose to the top became fabulously wealthy and politically powerful. They were able to control politicians, influence legislation, and sway government policy at every level. The gap between the rich and poor grew enormous, and the period became known as the Gilded Age.

At the beginning of the twentieth century, the tendency toward monopoly became so pervasive and business practices so egregious that the government finally intervened. When Teddy Roosevelt became president in 1901, he challenged the big banks and corporate giants and imposed government regulatory control over the excesses of capitalist monopolies. Anti-trust legislation was passed into law to prevent big business and the big banks from unfairly eliminating competition in the market place. But corporations constantly test the limits of the law, and search for ways to eliminate competition whenever possible. This is a battle that continues to this day.

Boom and bust

Another unfortunate feature of free market capitalism is that it is unstable. Economic activity regularly oscillates between periods of boom-and-bust. The market crash of 2008 was only the most recent in a series of business cycles that have occurred roughly every two decades for the past two centuries. The free market economy contains many positive feedback loops that make it subject to oscillatory behavior that swings up and down from economic expansion to recession and back. Most of these cycles are relatively small, but occasionally they are large, going from wild speculation to market crash. Typically, the crash is followed by months or even years of slow growth and high unemployment.

This boom-bust cycle is fundamentally rooted in the human emotions of greed and fear. The market value of stocks is based on what investors think they are worth, or more importantly, what they think they will be worth in the future. If investors think market values are going up, they will borrow money to buy, with confidence that they will be able to sell at a higher price, pay back their creditors, and make a profit. When the market is going up, investors get greedy. They see others getting rich, and don't want to be left behind. They borrow as much as possible and make overly optimistic investments. Consumers grow confident that their jobs are secure and incomes rising. They borrow as much as possible to buy cars, houses, furniture, and appliances on credit. The explosion of credit causes the economy to spiral upward until some of the investments fail to pay off, and consumers begin to max out their credit cards. At that point, market prices stop going up, and investments begin to fail. Investors cannot repay their creditors unless they sell their stocks. This drives stock value down.

When the market is going down, investors become fearful. Pessimism about the future causes investment to dry up. Stock values fall. Stocks bought on credit must be sold at a loss to cover loan payments. When loans cannot be repaid, investors go bankrupt and banks may fail. Fearful depositors try to get their money out of the banks. But the banks cannot give depositors their money because it is tied up in bad loans. Consumers stop spending, demand falls, profits turn into losses, and businesses lay off workers. When workers fear layoffs, they cut back on purchases. This leads to more layoffs. Unemployment rises, and the economy spirals downward.

This cycle is driven by human expectations that the future is going to be like the past. If the economy is going up, it will continue to go up. If it is going down, it will continue to go down. This is a common phenomenon in systems with feedback and time delays. It is well known in control theory as a limit cycle. It causes the economic system to overshoot, and oscillate above and below a long-term average growth rate that is sustained by technology growth.

During severe market crashes such as occurred in 1929, investment may drop to near zero. Investors will not invest if there is surplus inventory, unused productive capacity, and no consumer demand. Consumers cannot create demand without income. Businesses will not produce goods and services or hire workers unless there is consumer demand. So the economy can be stuck in depression for years.

Fiscal and monetary policies

There are two tools that the government has to smooth out the boom-and-bust cycle. These are fiscal and monetary policy. Fiscal policy is set by

Congress and executed by the president. It involves taxing, spending, and borrowing by the government. Monetary policy is set by the Federal Reserve System and the Treasury. Monetary policy controls the supply of money and credit, and determines interest rates.

Fiscal policies are heavily influenced by the theories of John Maynard Keynes, a British economist who advocated the use of fiscal and monetary measures to mitigate the adverse effects of economic recessions and depressions.[35] Monetary policies are most often associated with Milton Friedman, a professor at the University of Chicago and Nobel Prize winner in economics. Among scholars, he is best known for his theoretical and empirical research, especially monetary history and theory.[36] The experience of the Great Depression led Keynes to propose government fiscal stimulus as a means to get the economy moving again. Under Franklin Roosevelt, government spending for roads, government buildings, and national parks slowly revived the economy. But stimulus spending was controversial and resisted by political conservatives as wasteful and ineffective. Worries about excessive government debt created resistance to this kind of government spending, so fiscal stimulus programs were kept small relative to the size of the economy. It was not until the beginning of World War II that bipartisan support developed for massive government spending on manufacturing facilities to build tanks, trucks, ships, planes, guns, bullets, and bombs for World War II. This is what finally pulled the economy back to a place where private investors were willing to invest in a serious way.

The problem with fiscal policy is that it is subject to political pressures created by voters and political-action groups who want both lower taxes and more generous government programs for themselves and their friends and families. If the politicians fail to give the voters and interest groups what they want, they don't get reelected. Thus, it is very rare that fiscal policy produces a balanced budget even when the economy is growing rapidly. The tendency is to tax less, spend more, and make up the difference by borrowing, or in the last resort, by printing money. Borrowing to meet current operating expenses takes money away from productive investment in the short term, and generates unsecured debt that must be repaid with interest in the long term. Printing money is inflationary unless the newly created money is invested in

---

35   Keynes, John M. *The General Theory of Employment, Interest and Money.* Cambridge: Macmillan Cambridge University Press, 1936. Available online at http://www.marxists.org/reference/subject/economics/keynes/general-theory/. (accessed May 29, 2011).

36   "Marginal Utility of Money and Elasticities of Demand," *The Quarterly Journal of Economics,* Vol. 50, No. 3 (May, 1936), pp. 532–533.

productive assets, and even then it creates inflationary pressures in the short term before the investments pay back.

Inflation most often manifests itself when the economy is growing rapidly. During boom years, unemployment is low, workers are in demand, wages and salaries are rising, stock prices are going up, and everyone is doing well. Investors borrow money to invest in assets that are appreciating in value in anticipation that prices will continue going up. Consumers borrow money to buy luxury goods and services, and are seldom deterred by rising prices because they anticipate that their wages and salaries will continue going up. Cuts in government spending during an expansionary period are rare, because tax revenue is rising. Cuts in tax rates are much more common, which put more money into circulation and fuel both growth and inflation. As a result, fiscal policy can be used to stimulate an economy in recession, but is not an effective tool for controlling inflation.

This leaves monetary policy as the primary tool for controlling inflation. Monetary policy is set by the Federal Reserve Board, which is an independent body that is presumably not subject to political influence. But, monetary policy suffers from at least two problems that suggest that it may not be the best of all possible tools for controlling inflation.[37]

The first problem is that monetary policy works by limiting economic growth. Monetary policy stipulates that the cure for inflation is for the Fed to raise interest rates and tighten credit. This makes borrowing for investment more expensive. It lowers the rate of investment, slows productivity growth, and reduces the rate of economic growth. Eventually, inflation is brought under control—but at what cost? Slow economic growth means lost profits for business, lost jobs for workers, lost income for consumers, lost revenue for government, and slower rate of technological development, which is the ultimate source of prosperity.

A second problem with monetary policy is that interest rates have only a tangential and slow-acting effect on the consumer price index. It typically takes months for monetary policy to have a significant impact on consumer prices. It is well known from control theory that a time delay between measurement and action can produce instability, even in a system with negative feedback. The way to stabilize a system that has inherent time delay is to reduce the gain. As applied to monetary policy this means that economic growth is purposely limited to a rate that is stable, but far less than what is technically possible.

---

37   Keynes, John M. *General Theory of Employment, Interest, and Money*, Cambridge: Macmillan Cambridge University Press, 1936.

Limited access to credit for investment

A largely unheralded feature of capitalism is that investment banking is heavily biased in favor of those who are already rich. Few people can save enough to get rich by investing their savings. Most people become rich either by inheriting wealth or by borrowing money for investment. The most common way of growing rich is to borrow money, invest it in capital assets, pay off the loan with the return on investment, and reap the benefits in the form of dividends, interest, and rent. However, banks are reluctant to provide credit for investment to anyone that is not already rich. Any loan involves an element of risk that the recipient may not pay it back. To cover that risk, banks either require collateral or they charge high rates of interest.

Banks are only too willing to issue credit cards to the non-rich for consumer goods, such as food, clothing, vacations, and personal expenditures. Banks make money on credit cards because they charge interest rates that are high enough to cover the risk of default. Credit cards typically charge interest in excess of 14 percent on unpaid balances, and there are significant penalties for failure to pay on time. Banks will even loan money to consumers for homes because the loans are secured by the value of the real estate and the interest is subsidized by the government. But they lend money for capital investment only to those that already have capital assets to use as collateral. Loans for investment must carry much lower interest, because most safe investments pay returns of less than 10 percent.

As a result, the percentage of the population that derives most of its income from returns on capital investments is small. In fact, ours is not really a capitalist society. It is a worker society with a few capitalists who own most of the capital assets.

Unemployment

In the future, the biggest problem with capitalism may be unemployment. Until recently, businesses have needed large numbers of workers to produce the goods and services that society wants and needs. This created a prosperous middle class with good jobs that pay well. In the future, this may not be the case. Unemployment is already a significant and growing problem around the world, and wages have not kept up with inflation. In the recovery period after each recession, companies are slow to rehire workers laid off during the downturn. Companies discover that they can use technology and surplus capacity to increase output without hiring workers. So until market demand exceeds productive capacity, they don't hire. Of course, when people are out of work and paychecks are not growing, market demand declines, so recovery is slow.

In the future, advances in technology will generate productivity growth. Increased productivity means more output with the same or less input. The essence of productivity growth is that workers can produce more with less—less labor hours, less capital, and less raw materials. Productivity is good. Productivity is what enables us to live in homes that are warm in the winter and air conditioned in the summer, with glass windows and automatic dishwashing and laundry machines. Productivity is what enables us to drive cars, fly in airplanes, and ride in high-speed trains.

Improved productivity creates benefits that can be distributed in a number of ways. The options are:

- Lower prices in the market. When prices are lower in the market, consumers will buy more products and the market will expand.
- Increased quality of goods and services. When quality increases, consumers may shift preferences to higher-quality goods.
- Increased wages and salaries to the workers. When wages and salaries are increased, living standards for the workers will improve and companies may be able to attract better workers.
- Improved machines and equipment. When the company buys better machines and equipment, they may be able to achieve even greater productivity gains.
- Increased dividends to the stockholders. When the company distributes benefits to stockholders, the value of the company on the stock market will rise.

As technology improves exponentially, productivity grows exponentially. This is what separates us from stone-age lifestyle. As productivity increases, fewer hours of work are needed to produce the same amount or more of goods and services that people want and need. This trend began with the invention of stone tools and domestication of fire, and has progressed at a roughly exponential growth rate since. For millennia, progress was slow and uneven as empires rose and fell and knowledge was discovered and lost again. The pace of technology growth picked up sharply with the Renaissance and went into overdrive during the Industrial Revolution. With the deployment of each new technology, productivity increased. With each increase in productivity, the number of workers required to produce a given product has fallen. This trend has accelerated throughout the last 250 years, and the rate of growth today is faster than ever before.

The full effect of productivity growth generated by the Industrial Revolution was felt first in the agricultural sector. In 1790, 90 percent of the labor force was required to feed the population. Since that time, the introduction of technology into the agricultural sector has increased productivity of farm workers to the point where only 2.6 percent of the jobs are now required to feed the population. This is illustrated graphically in figure 3.1.

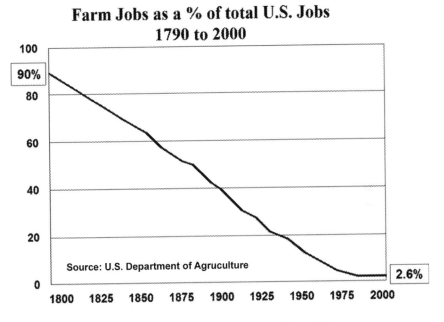

Figure 3.1. The decline in the number of farm workers needed to feed the population.

Since the invention of the cotton gin in 1793, innovations such as the steel plow, the McCormick reaper, the farm tractor, hybrid seeds, better fertilizers, and a continuous string of improvements in machinery and methods to prepare the ground, plant the crops, harvest the produce, and tend the animals have made it possible for fewer and fewer farm workers to provide food for the nation.

Fortunately, during most of that period, workers that were no longer needed on the farm were able to find jobs in the manufacturing industries. People migrated from the rural areas to the cities where factories were sprouting up, and where workers were needed for manufacturing jobs. Manufactured goods began to flood the market—electric lights, indoor plumbing, cotton clothing and sheets, furniture, dishes, washing machines, sewing machines, vacuum cleaners, automobiles, and roads to drive them on. Manufacturing jobs were needed to produce materials for railroads, ships, trucks, and construction of buildings, roads, bridges, and tunnels. Workers were needed in the construction, mining, transportation, and communication industries.

However, with the introduction of technology into manufacturing, the percent of the labor force required in manufacturing sector has experienced a similar decline as farming. As shown in figure 3.2, since the middle of the twentieth century, manufacturing productivity has risen, and fewer workers are required to produce the goods that the market demands.

Figure 3.2. percentage of jobs in manufacturing since WWII.

Following a spike during World War II, the number of jobs required to meet market demand for manufactured goods has declined as a percentage of the labor force from a peak of 37 percent at the end of WWII, to less than 9 percent today. Output has risen, quality has increased, while fewer and fewer workers are needed to produce the manufactured goods that consumers have money to buy.

In 2011, the number one priority of both political parties in the United States is jobs, jobs, jobs. In the current economic system jobs are critical for economic prosperity. When people lose their jobs, they lose their main source of income. They can't meet their mortgage payments. They can't afford medical expenses. They have to cut back on luxuries such as travel and vacation, eating out, going to the theater. They can't buy a new car, remodel their home, or buy new appliances. People who lose their jobs also lose their health insurance, so even a relatively simple illness can drive them into bankruptcy, and in some cases into homelessness. When large numbers of people lose their jobs, consumer demand falls, businesses lay off workers, and the economy spirals downward. Thus, it is no wonder that so much of our political energy is focused on creating jobs.

For most people, jobs are not only a primary source of income, but a status symbol. Jobs are a mark of identity, a source of pride, and a place where one can achieve. Most job environments are hierarchically structured. If you have a job, you have a boss, and you may have subordinates. Typically the higher in the job hierarchy, the bigger the salary and the greater the status. This organizational structure is quite efficient in management of large enterprises, and is widely employed in government, military, and corporate organizations. When you lose your place in the hierarchy, without finding another job that is equal or better in terms of pay and prestige, it can be frustrating and humiliating.

But corporations do not create jobs in order to provide income or social status for workers. Jobs are created when workers are needed to perform useful work in the production of goods and services. Sale of what is produced in the marketplace generates corporate income which goes to pay the employees, maintain the capital assets, pay rent, and procure raw materials. What remains are profits to be either reinvested in capital assets or distributed to the owners. Anything that can reduce the cost of labor, capital, rent, or raw materials goes directly into profits. So long as skilled and knowledgeable workers are required to produce the product, then workers can demand fair wages, and withhold their labor if they don't get them. However, as soon as workers are not needed, their jobs are eliminated and the workers are either reassigned or terminated.

Globalization

To some extent (the exact amount is debatable), the loss of manufacturing jobs in the United States has been caused by companies moving production facilities overseas. Transportation has become so efficient and inexpensive that it is economically feasible to mine ore in Wisconsin, cut timber in Brazil, and ship it to Asia or South America where it is turned into products that are returned to America to be sold. Manufacturing technology can be put anywhere in the world where labor is cheap. And once labor costs begin to rise, the technology can be moved again to where labor is cheaper still. This is a trend that will continue and even accelerate in the future as machines become more skilled, less expensive, and more reliable.

But America is not the only country experiencing job loss in the manufacturing sector. Economists at Alliance Capital Management discovered that twenty large international economies experienced the loss of

22 million manufacturing jobs between 1995 and 2002.[38] Even in China and India, technology is eliminating the need for workers in manufacturing.[39] As China modernized, it shut down or streamlined many inefficient state-owned enterprises. According to the World Bank, the Chinese laid off 43 million workers in the process of modernization between 1997 and 2004.[40]

As technology advances, productivity will increase, and fewer workers will be required to meet market demands. Simultaneously, as the human population grows, the number of people looking for work will grow. This virtually guarantees that unemployment will rise.

This trend could be reversed if the vast pool of poor people in the world were somehow turned into consumers. If that were done, more workers would be required to produce the goods and services that would be demanded by these new consumers. In some places, that is happening. In the rapidly growing economies of China, India, and South Korea, many of the poor are finding jobs and thus income. Newly employed workers receive wages and salaries that provide them purchasing power in the marketplace. This creates market demand for goods and services and creates jobs for workers needed for production. This creates customers for products and services and the demand for workers is growing. Millions of new jobs are being created. In rapidly growing economies, jobs are created for human workers required for the production of goods and services. Unemployment is typically not a big problem in a rapidly growing economy—at least not for now.

However, this will continue only until wages rise above the cost of automation. At that point, capital will replace labor in the production process. Wages and salaries will no longer keep pace with economic growth. As wages rise and the cost of automation falls, demand for human workers will taper off; and as the population rises, the supply of people looking for work will grow. The inevitable market result will be to put strong downward pressures on the price of human labor.

As human labor grows less necessary for the production of goods and services, the traditional methods developed by organized labor to force businesses to pay a significant percentage of their income to workers as wages and salaries will not be effective. The capitalist imperative is to minimize costs. To the corporation, wages and salaries are costs to be minimized.

---

38   Marinchek, John A. "Will American Manufacturing Jobs be Back?," *American Affairs* (March 23, 2010), http://americanaffairs.suite101.com/article.cfm/will-american-manufacturing-jobs-be-back#ixzz0tOm8V9Ia.

39   Reich, Robert B. "Manufacturing Jobs Are Never Coming Back," *Forbes.com*, May 25, 2009. http://www.forbes.com/2009/05/28/robert-reich-manufacturing-business-economy.html

40   Samuelson, Robert J. "China's stiff upper hand," *Washington Post Newspaper*, November 8, 2010, A15.

If goods and services can be produced and profitably sold without human workers, corporations will downsize. Workers that are no longer needed will be laid off. Corporate management's primary responsibility is to maximize return on investment to the stockholders (while paying themselves as much as they can). Laying off workers that are no longer needed is considered a mark of good management. When a large company announces it is going to lay off thousands of workers, its stock almost always goes up, and its managers receive bonuses.

In an article entitled "Manufacturing Jobs Are Never Coming Back," Robert B. Reich, the twenty-second Secretary of Labor and currently a professor at the University of California at Berkeley,[41] states that for the most part, things are made by machines, not people. Even China is eliminating factory jobs for the same reason the United States is eliminating them – because of increasing productivity. Reich argues that automation and technology are the reasons jobs are disappearing. Most significantly, he believes these jobs are never coming back.

Even after the economy recovers from recession to the point where companies again can profitably sell their products, not all the lost jobs will be refilled. As the economy recovers, employers discover that the newer technologies make it possible to operate the business with fewer workers. Much of the physical labor performed in manufacturing is already done by machines. Most jobs involve operating machines, managing other workers, or performing work that requires too much flexibility and hand-eye coordination to be automated. Humans are still needed to design, install, and repair machines, and perform mid-level management of the company. But as the machines get smarter, more efficient, and less costly, they will require fewer operators. As software becomes more intelligent, more efficient, and less costly, fewer managers will be required to run the business enterprise. As machines become more reliable, with built-in diagnostics and modular components, they will require less attention by maintenance workers. The essence of productivity improvement is that fewer workers can produce more goods and services at lower cost. The question that has loomed over the capitalist system since the days of the Luddite riots in 1812 is, what about those that are not needed for the production of goods and services? How are they supposed to get an income?

41   Reich, Robert B. "Manufacturing Jobs Are Never Coming Back," *Forbes.com,*
      May 25, 2009. http://www.forbes.com/2009/05/28/robert-reich-manufacturing-
      business-economy.html

New jobs

Of course, there will be new jobs for people that build intelligent machines, and program them with increasingly sophisticated software. There are many personal service jobs that cannot easily be performed by machines and cannot be shipped off shore. There will be new jobs in the health care industries as the population ages. There will be some new jobs in industries that support rapid rail and clean energy. The increased wealth generated by improved technology will enable companies to hire new workers—but only as many as are needed to generate profits for the owners.

The problem is that the number of new jobs created in high-tech industries will not be anywhere near the number of jobs lost in the traditional manufacturing and service sectors. Advanced degrees in electrical, mechanical, and software engineering will be required for the new high-tech jobs. But only a relatively small number of workers will be needed to run high-tech businesses and operate largely automated production facilities. Eventually, as machine intelligence grows, these jobs will be vulnerable as well. Information technology will be the next sector where intelligent systems technology will have an impact on productivity. Improved software will be much more user-friendly, easier to install, and simpler to maintain. Most of the software development process is already formalized. The need for low-level programmers is rapidly disappearing along with punched cards and paper tape readers. Future intelligent systems will be built with tools that improve productivity in software design, development, debugging, and testing. Automatic software-development systems will do most of the future programming tasks, as well as the training and learning tasks required for teaching machines how to adapt to changing conditions. Just because these activities have not yielded to automation in the past does not mean that they will not in the future.

The enabling technology is progressing exponentially. The computational power of the fastest supercomputers is already well into the range of the estimated computational power of the human brain. Within two or three decades, the computational power to perform most of the intellectual tasks required of middle management will reside on the average laptop available for purchase for a few hundred dollars over the Internet. Increasing knowledge of how the brain works, coupled with rapidly advancing techniques for building intelligent machines, will make possible dramatic improvements in computer systems that duplicate what most human workers do today at a fraction of the cost.

In the next two or three decades, scientists will likely discover how the brain actually works, and engineers will be able to build computer systems that are functionally equivalent to the human brain. The first simulations of entire brains have already been completed on the rat brain, and work is currently progressing on the cat brain. These are relatively simple models that do not address the higher-level functions of perception, world modeling, decision-making, and control of complex behavior. But progress is rapid and research programs are well funded. Before the middle of this century, it almost certainly will be possible for a computer to duplicate the cognitive capabilities of a well trained and highly skilled human. This will occur first on supercomputers in well funded research labs, but soon thereafter, will be available on laptops from WalMart and Best Buy. This will be a revolutionary turning point in human history. It will change everything. This is the Singularity envisioned by Ray Kurzweil.[42] It will drive a stake through what remains of the labor theory of value.

Most of the jobs that survive advances in intelligent machines will be in health care—doctors and nurses, and in personal services such as manicurists, masseuses, maids, and gardeners. The problem is that these jobs do not make things. They only provide services. Most of the material wealth generated by the future economy will be created by intelligent machines, not human workers, and will be distributed as income to the owners through return on investments, not to workers as wages and salaries.

Slow economic growth

The rate of economic growth of the current capitalist economy is far beneath what is physically and technologically possible. This can be seen from the performance of the US economy in the years between 1939 and 1945. During that period the US economy grew at a rate of more than 14 percent per year. This transformed the US economy from a basket case in 1939 to the world's dominative economic superpower in 1945. Real GDP grew by 188 percent in six years, from $950 billion in 1939 to $1786 billion in 1945 (in 2000 dollars.)[43] In less than a decade, the United States catapulted from near bankruptcy to the world's richest creditor. And this was accomplished while taking 12 million able-bodied young men out of the labor force to serve in the military, and destroying most of the productive output on the battlefield. Other examples of rapid annual growth are the Japanese economy, which grew at 10.5 percent from 1950 to 1973,[44] the Chinese economy, which has averaged more than 9 percent growth since 1978, and the "Asian tigers"— Hong Kong, South Korea, Singapore, and Taiwan—that have had an average

---

42   Kurzweil, Ray. *The Singularity Is Near*, Penguin, New York, 2005. http://www
       .singularity.com/.

43   P. Krugman, and R. Wells, 2009, *Macroeconomics, 2nd Edition*, Worth Publishers,
       New York, p. M-1.

44   P. Kennedy, 1987, *Rise and Fall of the Great Powers*, p. 417.

growth rate well in excess of 7 percent over the last 15 years. In several peak years, the Chinese economy grew more than 13 percent and per capita income in China has nearly quadrupled in the last fifteen years.[45] Average citizens of South Korea are 638 percent richer today than their parents were in 1960.[46] Clearly, technologically advanced economic systems are capable of sustaining annual growth well in excess of 3 percent for decades.

However, despite a long series of technological advances in manufacturing, transportation, and communications that have made industry much more efficient and productive, monetary policy has limited the US average annual growth rate to 3.9 percent since the end of WWII. After WWII, US investment policies reversed course. Instead of high rates of saving and investment, US fiscal and monetary policies switched to high consumption and low savings. Over the past sixty years, US investment in capital assets has consistently lagged the world average.[47] Meanwhile, an increasing proportion of the population has moved from industrial jobs to service jobs that have much lower rates of productivity growth. [48] [49]

The currently accepted economic wisdom is that the maximum sustainable rate of economic growth is 3 percent per year. Anything higher is considered "unsustainable" because it will drive unemployment down to the point where workers can demand higher wages and thereby cause inflation. The Congressional Budget Office predicts that the US economy will grow at an average of 3 percent per year for the next decade. [50]

This suggests that the current fiscal and monetary policies may not be the best of all possible tools for simultaneously producing rapid economic growth and low inflation.

---

45   International Monetary Fund data, "Why Is China Growing So Fast?" Zuliu Hu and Mohsin S. Khan, http://www.imf.org/EXTERNAL/PUBS/FT/ISSUES8/INDEX.HTM.

46   International Monetary Fund, "Growth in East Asia: What We Can and What We Cannot Infer," Michael Sarel, http://www.imf.org/external/pubs/ft/issues1/.

47   World Bank data for US investment relative to the world, http://www.google.com/publicdata/explore?ds=d5bncppjof8f9_&ctype=l&strail=false&nselm=h&m et_y=ne_gdi_totl_zs&hl=en&dl=en#ctype=l&strail=false&nselm=h&met_y=ne_gdi_totl_zs&scale_y=lin&ind_y=false&rdim=country&idim=country:USA&tdim=true&hl=en&dl=en.

48   L. Thurow, 1996, "America among equals," in S.J Unger (ed.), *Estrangement: America and the World*, Oxford University Press, New York.

49   P. Kennedy, 1987, *Rise and Fall of the Great Powers*, p. 433.

50   Congressional Budget Office, http://www.cbo.gov/ftpdocs/108xx/doc10871/BudgetOutlook2010_Jan.cfm.

We Could Do Better

While free market capitalism is the best system yet devised to generate wealth, raise efficiency, and create prosperity, it is far from perfect. It has a number of undesirable features. Some of these have been mitigated by laws giving workers the right to form unions, negotiate compensation, and gain protection from unsafe working conditions. Other defects have been corrected by government regulation of the marketplace, insurance of bank deposits, and a graduated tax system. The current system is the best yet, but far from the best that could be. Particularly at this point in history when capital is replacing labor as the principal factor in the production of wealth, modifications are needed to deal with the impact of continued productivity growth on employment.

In the next chapter I will suggest how the capitalist system could be improved by boosting the rate of investment, broadening the ownership of capital, and gradually shifting from wages and salaries to dividends and interest as the primary mechanisms for distributing wealth.

CHAPTER 4 |||||||||||||||||||||||||||||■➡

Peoples' Capitalism

PEOPLES' CAPITALISM[51] IS A PLAN to broaden the ownership of capital to include the entire citizenry. The ultimate goal of Peoples' Capitalism is to create the vision presented in chapter 2. In this vision, everyone would have enough income from ownership of capital assets to live comfortably without having to work. In this vision, there would be plenty of interesting and well-paying jobs available for those who wish to work to earn additional income, to contribute to society, or to achieve social status from success in the working world. But no one would have to work to live. People would be free to seek higher education, to travel, to homestead, to volunteer, to pursue a hobby, or to start a business. The disabled and elderly would be financially secure.

Peoples' Capitalism is grounded on the same fundamental principles as free market capitalism. The means of production and distribution would continue to be privately or corporately owned and operated for profit. Individuals and firms would continue to have the right to earn income and to purchase labor for wages. Corporations would continue to compete for customers in free markets with open competition. Regulation would be achieved mainly through the operation of market forces where prices and profits dictate where and how resources are used and allocated. Government intervention would be limited to funding research and development, policing markets, and prosecuting fraud. Ambitious, talented, and hardworking individuals would have ample opportunity to become extremely wealthy.

---

51   Albus, James S. *Peoples' Capitalism: The Economics of the Robot Revolution*, Kensington, MD: New World Books,1976. Available online at http://www. PeoplesCapitalism.org.

The primary mechanism for broadening the ownership of capital would be to provide credit to every citizen for investing in capital assets. Restrictions would be placed on this credit to assure that it be invested in capital assets that have a high probability of providing sufficient returns to pay back the investments and provide income for the individual investors. By this means, everyone would have an opportunity to acquire a growing ownership share in the means of production, and to derive a growing income stream from ownership of capital assets. In the early phase, the income stream from capital ownership would be small, because investment portfolios would be small. However, as portfolios grow year by year, the income from ownership of capital assets would grow also. Within two decades, return on investment would provide a significant supplemental source of income for working families. Within four decades, ownership of capital could become the primary source of income for most people. At every step of the way, there would be plenty of room at the top for entrepreneurs, corporate executives, venture capitalists, investment bankers, doctors, lawyers, and entertainers. But there would also be a rising floor at the bottom that would prevent anyone from sinking into poverty.

Peoples' Capitalism Background

Peoples' Capitalism is an original concept independently developed by James S. Albus and first published in 1976 (*Peoples' Capitalism: The Economics of the Robot Revolution*). Available online at the Peoples' Capitalism website http://www.PeoplesCapitalism.org, this book extends the original concept with a quantitative model that predicts the potential effect on the future economic system.

The fundamental idea of distributing profits from industry to the general public was first proposed by Major Clifford Hugh Douglas, a British engineer and pioneer of the Social Credit economic reform movement. Between 1916 and 1920, he developed his economic ideas and published two books in 1920 (*Economic Democracy* and *Credit-Power and Democracy*). Douglas collected data from over a hundred large British businesses and found that in every case, except that of companies heading for bankruptcy, the sums paid out in salaries, wages, and dividends were always less than the total costs of goods and services produced each week: the workers were not paid enough to buy back what they had made. Freeing workers from this system by bringing purchasing power in line with production became the basis of Douglas's reform ideas that became known as Social Credit. There were two main elements to Douglas's reform program: a **National Dividend** to distribute money (debt-free credit) equally to all citizens, over and above their earnings, to help bridge the gap between purchasing power and prices; also a price adjustment mechanism, called the **Just Price**, which would forestall any possibility of inflation. The Just Price was designed to reduce retail prices by a percentage that reflected the physical efficiency of the production system. Douglas observed that the cost of production is consumption; meaning the exact physical cost of production is the total resources consumed in the production process. As the physical efficiency of production increases, the Just Price mechanism will reduce the price of products for the consumer. Unfortunately, the effect of the Just Price mechanism is to reduce profits to zero, and hence to eliminate incentives for investment.

Over the years, Social Credit political movements developed in Great Britain in the form of the British People's Party; Social Credit Party of Great Britain and Northern Ireland; in Canada in the form of Abolitionist Party of Canada; British Columbia Social Credit Party; Canada Party; Christian Credit Party; Committee on Monetary and Economic Reform; Les Démocrates; Manitoba Social Credit Party; New Democracy; Pilgrims of Saint Michael; Ralliement créditiste; Ralliement créditiste du Québec; Social Credit Board; Social Credit Party of Alberta; Social Credit Party of Canada; Social Credit Party of Ontario; Social Credit Party of Saskatchewan; and in Oceania Australian League of Rights; Douglas Credit Party; New Zealand Democratic Party for Social Credit; Social Credit Party; Solomon Islands Social Credit Party.

The Social Credit party of Alberta elected its leader to Premier in 1935 and tried to implement social credit by issuing "prosperity certificates" to Albertans. However, this measure was disallowed by the Supreme Court of Canada on the basis that only the federal government of Canada was authorized to issue currency. The Social Credit Party of Alberta formed nine consecutive majority governments spanning thirty six years, one of the longest spans of a single party in government in Canadian history.[52] Elsewhere, Social Credit parties never exceeded 16 percent of the vote.

In 1958, Louis Kelso and Mortimer Adler published another version of the basic idea as the *New Capitalist Manifesto*. In 1967, Louis Kelso and Patricia Hetter published the *Two Factor Theory*. Based on this theory, legislation enabling Employee Stock Ownership Programs (ESOPs) was enacted into law by the U. S. Congress.

Peoples' Capitalism extends the stock ownership concept to every citizen, and would diversify the ownership portfolio from a single company to a broad range of investment opportunities.

Although it was advanced under several names including "Universal

---

52   http://en.wikipedia.org/wiki/Canadian_social_credit_movement#Alberta

Capitalism," "The Second Income Plan," and "Two-Factor Theory," in his later writings, Louis Kelso settled on the term "binary economics" as the name of his theory.

Over the past thirty years, Norman Kurland has been active in the promotion of binary economics. Kurland founded the *Center for Economic and Social Justice* to promote the notion of widespread capital ownership. *Capital Homesteading,* published by the Center, suggests detailed mechanisms by which access to credit for capital investment can be made available to all (Kurland et al 2004).

Scholarly support for the concept of access to investment capital for all is found in Ashford and Shakespeare (1999), and in Ashford (1998, 1996, 1990). In academia, Professor Ashford has been the leading proponent and explicator of binary economics. In his writings, he has distilled the unique features of binary economics to three fundamental principles: (1) labor and capital are independent (or binary) variables in production; (2) the more broadly capital is acquired the greater the market incentives to profitably employ existing capacity and invest in additional productive capacity; and (3) everyone should have the right to acquire capital with the earnings of capital.

A unique feature of Peoples' Capitalism is the mandatory savings plan. The Personal Savings Plan withholding enables sufficient investment capital to be inserted into the economy to generate economic growth rates of 6 percent to 9 percent. The Employee Stock Ownership proposals of Adler, Kelso, and Kurland are much more modest in scope, and benefit only the employees of single companies. Only if there is a new mechanism, such as the PSP for combating inflation, could the nation's investment rate be increased by enough to achieve the results envisioned by Peoples' Capitalism.

Those that have wealth would keep it, and would have ample opportunity to earn more. The disparity between the top and bottom might even grow. However, Peoples' Capitalism would raise the floor. Those at the bottom would no longer be poor. Everyone would have a minimum income from ownership of capital assets that would make them financially secure and able to live comfortably. Peoples' Capitalism would function by giving every citizen meaningful access to credit for investing in capital assets. This would be achieved not by reducing access to credit for investment to those who already have it, but by creating a new line of credit for those who do not have it now. In this way, everyone would have the ability to build a substantial portfolio of capital assets that would provide a supplemental source of income from ownership of the means of production.

Under Peoples' Capitalism, economic policy would have two new features:

1) Access to credit for investment would be available to everyone.

2) Savings would replace monetary restraint as the primary tool for controlling inflation.

Access to Credit for Investment

The financial engine of capitalism is the ability to borrow money for investing in capital assets (e.g., factories, machinery, office buildings, shopping centers, housing developments, corporate stocks and bonds). Successful investments are those that pay dividends, interest, or rent in an amount sufficient to pay off loans from creditors and return a profit to the investors. Without the ability to borrow money for investment, capitalism would not work. Large projects such as building a factory, a chemical plant, a power plant, a railroad, an airline, a shipping company, a public utility, a hotel, a restaurant, a resort, or a casino all depend on the ability to borrow money. Under the current system, only rich individuals, big corporations, governments, and financial institutions have access to credit for investing in capital assets. Banks make loans for investment only to those who already have capital assets that can be used for collateral. As a result, under the current system, the owners of capital assets represent a small percentage of the population. Peoples' Capitalism is designed to modify the rules, not to punish the current owners of capital, but simply to provide access to credit for purchase of capital assets so that anyone can become a capitalist. Peoples' Capitalism would not level the playing field, but it would broaden it. It would let ordinary people get into the game by giving them meaningful access to credit for acquiring capital assets.

A simplified diagram of the proposed Personal Investment Plan (PIP) is illustrated in Figure 3.1. The central bank (in the US, the Federal Reserve) would open its discount window and issue credit to local banks for making self-liquidating loans to individual citizens for investment in approved mutual funds. The bank loans to individuals would be secured by shares in the mutual funds, and the loans would be repaid with dividends earned by the shares. Whatever residual dividends are left after monthly payments are met would become personal income for individual investors. Once the loans are repaid in full, the shares would become the property of the individual investors. From then on, all of the dividends from the investments would flow directly to the individual citizen investors. Details of possible mechanisms for implementing the PIP are described in Kurland and associates' *Capital Homesteading for Every Citizen*.[53]

All investments by individual citizens using money from the PIP would

---

53   Kurland, Normand, Dawn K. Brohawn, and Michael D. Greaney. *Capital Homesteading for Every Citizen*, Washington, DC: Economic Justice Media, 2004. See also the Center for Economic and Social Justice at http://www.cesj.org/.

have to be made in approved mutual funds. To get on the approved list, a mutual fund would have to have a proven track record of success in producing return on investment. Approved funds would be frequently audited by the Securities and Exchange Commission, the Federal Reserve, and appropriate committees of Congress, with publicly published results. Stock in the approved mutual funds would have to trade on the open market, and funds that fail to score among the top funds would be removed from the approved list. A good example of this form of investment is the Thrift Savings Plan available to government workers.[54] Many 401K plans would qualify. Under the PIP, individual investors would choose from a menu of different types of investments including Treasury notes, bonds, stocks, or other securities, including those sold on foreign exchanges. Individual investors would have a choice of risk vs. return in their individual mix of investments, similar to the choice available under the Thrift Savings Plan.

The credit issued by the Federal Reserve to the banks would be at zero interest, but loans from the banks to individuals would carry up to a 2 percent service charge to cover administrative costs and insurance against catastrophic losses. These loans would have a payback period of thirty years. Insurance charges would vary depending on the amount of insurance coverage, and the level of risk in the individual's portfolio.

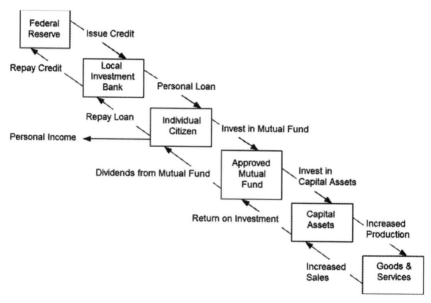

Figure 4.1. The Personal Investment Plan (PIP) for the flow of investment money under Peoples' Capitalism.

---

54   The US Government Thrift Savings Plan website https://www.tsp.gov/index.shtml.

Increasing Investment

Everyone agrees that increasing investment would be a good thing and should be encouraged. Investment is an expression of confidence in the future. It is well established that investment stimulates economic growth, encourages technological innovation, and increases productivity. These are all things that should be encouraged. Both conservatives and liberals are in favor of increasing investment. They differ only in the means for achieving increased investment.

Conservatives argue that cutting taxes is the best way to stimulate investment. The presumption is that if individuals have more income, they will invest more in capital assets. Unfortunately, most taxpayers spend most of their income on consumer goods rather than investing it. While the rich are more likely to invest than the poor, the rich spend a considerable fraction of their discretionary income on yachts, multiple homes, designer clothes, and expensive cars. In reality, only a small fraction of tax cuts, even to the rich, end up as investments in capital assets.

Liberals argue that government spending on roads and bridges are investments. They consider education and health care to be investments. They believe research and development are investments. They think that expenditures for public health and safety are investments. But all government spending is not investment. For example, Social Security transfer payments are not investments. Most recipients of Social Security payments use them almost entirely for consumption. Farm subsidies are not investments except to the extent that they insure against food shortages. Defense spending is not an investment except in so far as it builds industrial capacity, defends our economic interests around the world, and protects our security here at home. Much of what is spent on health care is not an investment. About one-third of Medicare and Medicaid expenditures are for the last year of life where return on investment is not a consideration.[55]

The Peoples' Capitalism PIP differs from both conservative and liberal approaches. It would directly increase investment in capital assets by making credit available to individual citizens for purchase of shares in investment funds that invest in businesses that earn profits and pay dividends. None of the money would be spent for consumer goods. None would be spent by the government for any purpose. All of the money goes directly into the hands of professional investment managers with strong incentives to generate long-term profits for their shareholders. All of the money goes to

---

55   Hogan, Christopher, June Lunney, Jon Gabel, and Joanne Lynn."Medicare Beneficiaries' Costs of Care in the Last Year of Life," *Health Affairs* 20, no. 4 (July 2001): 188-195 http://content.healthaffairs.org/cgi/content/full/20/4/188.

increase productive capacity for goods and services that satisfy consumer demand in the marketplace. Dividends earned on investments would be used to pay back the loans, and the residual would provide income directly to the individuals that made the investments. This allows investment to be increased in a manner such that everyone benefits individually and directly from increased productivity and faster economic growth.

Investing borrowed money is how capitalists become rich. Few people become rich by saving their wages or salaries and putting them into saving accounts. Most capitalists achieve riches by borrowing money, investing it, and using the return on investment to pay back their loans. The Peoples' Capitalism PIP would make it possible for average folks to participate in the capitalist game.

How much investment is optimal?

There are many who believe that the current rate of investment is optimal. They argue that if more investment could usefully be made, the market would provide it. This brings to mind the arguments made by Dr. Pangloss in Voltaire's *Candide*.[56] Pangloss maintained, against all contrary evidence of war, famine, earthquakes, disease, brutality, and slavery, that "this is the best of all possible worlds," because if the world could be any better, God would have made it so, because he is a benevolent deity. Present-day Panglossian economists maintain that the current level of investment must be optimal, because if it weren't, the market would make it so, because of the "invisible hand."

Yet, there are many reasons, both theoretical and empirical, to believe that the current rate of investment in the United States is far below optimal. The primary cause of recessions is that investors slow down their rate of investing. If the rate of investment were always optimal, it would not fluctuate erratically. Both political parties constantly tout policies designed to increase investment. The now infamous bank bailouts of 2008 and 2009 were designed to enable banks to loan money for investment. A great deal of the anger against the bailouts is because the banks failed to use the money for investing.

One can argue that even under normal times the investment rate of the United States is consistently suboptimal. Over the past sixty years, US investment in capital assets has consistently lagged the world average by more than 2 percent of GDP,[57] and since 1970 has fallen far behind the Chinese. In 2007, the US

---

56   Voltaire, Candide (1759). Literature.org, http://www.literature.org/authors/voltaire/candide/

57   World Bank data available online at http://www.google.com/publicdata/explore?ds=d5bncppjof8f9_&ctype=l&strail=false&nselm=h&met_y=ne_gdi_totl_zs&hl=en&dl=en#ctype=l&strail=false&nselm=h&met_y=ne_gdi_totl_zs&scale_y=lin&ind_y=false&rdim=country&idim=country:USA&tdim=true&hl=en&dl=en.

gross capital formation rate was 18.3 percent of GDP, while the Chinese rate was 44.4 percent.[58] Between 1960 and 1972, when Japan's economy was growing at an average rate of 10 percent, the Japanese gross capital formation rate averaged 35 percent. Since 1980, Japan's investment has fallen sharply, and in 2007 it was 24.1 percent.[59] Over that same period the Japanese economic growth rate dropped from 10 percent into recession, and today is below that of the United States.[60] The lesson is that rapid economic growth derives from a high rate of investment. If any country wants the benefits of a rapid economic growth rate, it needs to commit considerably more than 20 percent of GNP to investment.

One problem with the current system is that the classical tools of fiscal and monetary policy are not adequate to induce a significant increase in the investment rate. By tradition, the American political system does not permit the government to directly invest in private industry for the purpose of making a profit. As a matter of policy, government can invest in public infrastructure that supports industry, such as roads and bridges, public schools, and research, but not in shares of private companies. (This explains why the 2009 TARP program that rescued the automobile industry was so controversial.) The government can also encourage investment by giving tax cuts to private businesses and individuals in the hope that they will invest. But, tax cuts do not guarantee that the money will actually be invested in productive assets. The Federal Reserve can encourage the private sector to save by raising interest rates, but that raises the cost of borrowing money which discourages investment.

Another problem is that private investors will almost always invest less than what is optimal for the economy as a whole. Private investors are motivated by private profits, not the public good. It is well known that private investors cannot capture all the benefits of their investments. There are many benefits that accrue to other companies and the public in general that are not reflected in the return on investment that private investors realize. A classic study by economist Edwin Mansfield in 1977 found that the median social rate of return for seventeen industrial R&D projects was more than double the median private rate of return. The median social rate of return was 56 percent whereas the median private rate of return was only 25 percent.[61]

58   World Bank data available online at http://www.google.com/publicdata/explore?ds=d5bncppjof8f9_&ctype=l&strail=false&nselm=h&met_y=ne_gdi_totl_zs&hl=en&dl=en#ctype=l&strail=false&nselm=h&met_y=ne_gdi_totl_zs&scale_y=lin&ind_y=false&rdim=country&idim=country:USA:CHN&tdim=true&hl=en&dl=en.

59   World Bank data available online at http://www.google.com/publicdata/explore?ds=d5bncppjof8f9_&ctype=l&strail=false&nselm=h&met_y=ne_gdi_totl_zs&hl=en&dl=en#ctype=l&strail=false&nselm=h&met_y=ne_gdi_totl_zs&scale_y=lin&ind_y=false&rdim=country&idim=country:USA:CHN:JPN&tdim=true&hl=en&dl=en.

60   World Bank data available online at http://www.google.com/publicdata/directory.

61   Mansfield, E., et al., "Social and Private Rates of Return from Industrial

This is because not all the benefits of an investment can be captured by the companies making the investments. Even when the company has monopoly rights granted by patents, some of the benefits of technological innovation "spill over" to other companies, and to society as a whole as people employ the improved products that were generated by the investment.

More recently, Link and Scott reported that, on average across eight generic information technology R&D projects studied, the companies investing in R&D received only 13.5 percent of the profits generated by all firms from products developed and knowledge discovered as a result of the R&D.[62] Many companies that benefited from the results of the R&D contributed nothing to the investment. In a sample of fourteen Small Business Innovation Research (SBIR) Program R&D projects covering a range of technologies, Scott found that, on average, small businesses investing in R&D received only 30.5 percent of the profits created by their R&D results.[63] It should be noted that in all of these studies, the definition of "social return" includes only the profits earned by all industries, <u>not</u> the benefits to society as a whole. The benefits of investment to society as a whole thus exceed the so-called social returns of these studies, which far exceed the returns captured by the private individuals or companies making the investments. This means that market incentives for private investors will systematically produce less investment than is optimal for society as a whole. Thus, an increase in the rate of investment above that provided by the private market is economically justified for society as a whole.

Of course, at any level of investment, investors pick the most lucrative investments. Thus, it can be argued on the basis of diminishing returns that if the rate of investment is increased, the average rate of return will be somewhat lower than before. However, diminishing returns to capital has not occurred. Over the past one hundred years, the returns to capital have been quite constant, except for a step function that occurred during WWII when the returns to capital jumped from about .25 to .3. This was because of technological progress. Increased investment generates increased productivity because it enables innovation.

Innovations," *The Quarterly Journal of Economics*, vol. 91, no. 2, (May 1977): 221–240.

62   Link, Albert N. and John T. Scott. "Public/Private Partnerships: Stimulating Competition in a Dynamic Market," *International Journal of Industrial Organization*, Volume 19, issue 5, (April 2001) 763–794.

63   Scott, J. (2000) "An Assessment of the Small Business Innovation Research Program in New England: Fast Track Compared with Non-Fast Track Projects," in *The Small Business Innovation Research Program: An Assessment of the Department of Defense Fast Track Initiative*, Edited by Charles W. Wessner, Washington, DC: National Academy Press, 104–140.

A higher rate of investment would enable companies to buy new machines and modernize their production facilities. It would stimulate companies to innovate and upgrade their business practices. It would encourage them to train their workers to be more efficient. It would generate higher productivity that would shift the production-possibility frontier so that rising productivity overcomes the law of diminishing returns. Given sufficiently high rates of investment, the rate of return on capital investment may actually accelerate like it did during WWII.

How much should investment be increased?

How much credit should the Federal Reserve issue to banks for loans to individuals for investment? The studies by Mansfield and Scott suggest that the optimal rate of investment may be twice the current rate. In 2008, GDP was $14.6 trillion. In 2007, the US capital formation rate was 18.3 percent of GDP. Doubling the rate of investment would require an additional 18.3 percent of GDP, or $2.6 trillion of new investment. The resulting investment rate of 36.6 percent would still be less than the 2007 Chinese investment rate of 44.4 percent, but represents such a large amount as to strain credulity. An increase of this magnitude may turn out to be what we would consider desirable when we understand the possibilities that lie before us. However, with a nod to practicality, I will first examine a less dramatic proposal for an increase of 10 percent of GDP. Ten percent of GDP is $1.46 trillion (or about twice the Obama stimulus package). Given that the US gross capital formation rate has averaged more than 2 percent lower than the world average since at least 1960,[64] and is more than 25 percent of GDP less than the Chinese investment rate, an increase of 10 percent of GDP seems reasonable. If the Peoples' Capitalism model holds up, increasing the investment rate by 18 percent at some time in the future might seem feasible, even compelling.

If the Federal Reserve were to issue enough credit to increase the investment rate by 10 percent of GDP, the first and almost immediate result would be to stimulate the economy to grow by at least the amount of newly created money issued by the Federal Reserve for investment minus the amount of savings withheld to prevent inflation. The stimulus effect would be amplified by a multiplication factor, because each dollar inserted into the economy gets spent more than once in a year. If we assume a very conservative multiplication factor of say 1.2, then 10 percent of GDP new money inserted into the economy would produce a 12 percent increase in the GDP. To the extent that this increase would absorb the surplus productive capacity of the current economy, it would not be inflationary, and no PSP mandatory saving

---

would be required. In 2010, the official unemployment rate was 9.5 percent, and if we count those that have stopped looking for work and those that are involuntarily working less than full-time, it may have been as high as 15 percent. Thus, an increase of 12 percent in the economic growth rate might be achieved largely without inflation. In any case, it would quickly pull the economy out of recession and reduce unemployment to a low level.

Any increase in productivity generated by the increased investment rate would act to reduce the inflationary effects of the PIP investment increase. It is widely accepted that increasing the investment rate generates about 0.1 percent increase in productivity for every 1 percent increase in the investment rate simply through capital deepening.[65] Samuelson provides data suggesting that productivity growth from technological progress over the past one hundred years has grown on the order of 3.8 times faster than the increase due to capital deepening.[66] This suggests that each 1 percent increase in the investment rate would produce .38 percent increase in productivity.

Thus, a 10 percent increase in the investment rate could produce as much as a 3.8 percent increase in the productivity growth rate over and above current value of 2.3 percent. The result would be a total factor productivity growth rate of a bit more than 6 percent. This productivity growth would subtract from the inflationary effect of injecting 10 percent of GDP new money into the economy, producing only a 4 percent inflationary push. Given the current underutilization of more than 12 percent of the work force, it is entirely possible that a 10 percent increase in the investment rate might produce no inflationary impact at all, at least not until the economy recovers from recession.

Of course, once inflation does become a problem, PSP mandatory savings would kick in. PSP mandatory savings would temporarily reduce consumer disposable income by the amount needed to keep demand in balance with supply, and thus prevent inflation.

Once the economy recovers and unemployment falls to less than 4 percent, continuation of the 10 percent increase in investment would require additional PSP savings. Assuming a 6 percent rate of growth in total factor productivity, the inflationary effect would be about 6 percent. To counteract this would require a 6 percent PSP mandatory savings rate. Of course, by

---

65 Capital deepening is an increase in the total amount of capital at a given level of technology. Each 1 percent increase in the amount of capital typically produces .33 percent increase in productivity, and the amount of capital is roughly equal to 3 times the GDP. Thus, each 1 percent of GDP increase in investment generates about .3 percent growth in the amount of capital. which generates .1 percent increase in productivity. For details, see Paul Krugman and Robin Wells, *Macroeconomics,* p. 234.

66 Samuelson, Paul and William Nordhaus. *Economics, 13th edition*, New York: McGraw-Hill, 1989, 863–4.

diverting consumer disposable income into temporary savings, the PSP would reduce economic growth by the amount of the savings. A 6 percent savings rate would reduce the economic growth rate by 6 percent, so the net result would be a real economic growth rate of 6 percent and an increase in the personal savings rate of 6 percent.

The Problem of Inflation

The creation of credit by the central bank to finance loans to individuals for investment would effectively create money. This is inflationary. In general, issuing a loan of any kind increases the supply of money, and the repayment of any loan reduces the supply of money. Money is an artificial construct for the purpose of facilitating commerce and storing wealth. Prices are set by the ratio of money to wealth. What things cost is determined in the market by the ratio between the supply of money available to consumers to spend and the supply of goods and services available for sale. If the bank issues credit, it increases the supply of money. If that money goes into investment, it creates demand for raw materials, new plants and equipment, and workers to operate the businesses. Until new workers, plants, and equipment produce more goods and services, the credit issued by the bank will cause the price of materials, machinery, and labor to rise. However, as soon as the new products and services begin to arrive on the market, prices begin to fall. Once the investments are profitable, the inflationary effect goes away as the supply of goods and services increases. Dividends from the investments provide income to consumers to buy what is produced by new productive capacity. In the end, both supply and demand rise together, and prices remain stable.

Figure 4.2 illustrates the time line for a typical investment.

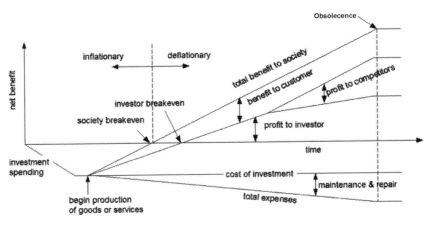

Figure 4.2. A time line of the costs and benefits of investment.

At time = 0, when credit is issued for investment, money is created and investment spending begins. This produces a need for architects and engineers to design plants, factories, and equipment. Construction workers are needed to build buildings and facilities. Programmers must be hired to write code for information and control systems. Workers get paid before the new plants go into production, and their income translates into increased demand in the market before there is any change in the supply of goods and services. This causes short-term inflation. However, once a new supply of goods and services begins to enter the market, positive cash flow begins. If all goes well, the positive cash flow eventually becomes greater than the investment money spent, loans are repaid out of return on investment, and the investor makes a profit.

Investment also generates productivity increases, because it introduces new and better technology into the production process. This means that more wealth can be produced at lower cost (i.e., less raw materials, more efficient machinery, and/or fewer workers). Thus, investment spending is inflationary only in the short term. In the long term, investment spending creates more wealth at lower cost. This is the ultimate source of real economic growth.

Thus, in the long term, creating credit for investment is not inflationary. However, in the short term it is. The problem lies in the time delay between when investment spending begins, and the social breakeven point is reached. In the period before positive cash flow, there is little inflationary difference between borrowed money spent on consumer goods and borrowed money spent on capital assets. During this early period, demand exceeds supply because purchasing power finds its way into the market before any new products and services appear in the market.

However, after production starts, positive cash flow begins and the inflationary effect of investment grows smaller until the social breakeven point when the inflationary impact is reduced to zero. After the social breakeven point, the benefit to society in terms of new products and services exceeds the cost of the investment, and investment becomes deflationary. At that point, if dividends are paid to investors, they have more money to spend, and demand increases to match the increase in supply. There are more goods and services in the market, and consumers have more money to spend. Supply tracks demand and prices stabilize at a higher level of output.

Note that society as a whole reaps benefits before and beyond what the private investor is able to capture. The social benefits of investment are not simply monetary. Investment in research and development increases technology in general. This may lead to new methods of production that are more efficient, less costly, and more environmentally friendly in other

60

industries as well. Thus, the social breakeven point occurs before the investor breakeven point, and in the long run, everyone benefits. But the social benefits always exceed the benefits to the private investor—and sometimes by a lot.

Eventually, the gain to the investor levels off as competitors enter the market, patents expire, and new products are introduced that make the investor's product obsolete. This adds to the risk that investors may never recoup their money. Investment is a risky business, and at best, individual investors risk more and gain less from their investments than society as a whole. It is therefore not surprising that private investors systematically invest less than what would be optimal for society as a whole.

Nevertheless, private investors can and do profit from good investments. Wise investments in new productive capacity generate new products and services that can be sold in the market for a profit. In the process, jobs are created that distribute income to workers. Income to workers allows them to become customers for the new products and services. The net result is profits for owners and income for consumers. The economy grows and the standard of living increases. This is the genius of capitalism. It is why capitalism is the most productive economic system ever devised.

However, as productivity grows and markets become saturated, fewer and fewer workers are needed in the production process. In this case, unemployment grows, total worker income falls, and there are fewer customers to buy what is produced. This is a situation that the current model of capitalism does not adequately address. It is the problem that Peoples' Capitalism is designed to solve.

Tools for controlling inflation

As was discussed in chapter 3, the current policy prescription for controlling inflation is for the Federal Reserve to raise interest rates and tighten the money supply. Unfortunately, this is largely counterproductive. Investment is the primary engine of productivity growth which is the ultimate source of economic growth. Policies of monetary restraint designed to control inflation have the side effect of reducing investment, slowing economic growth, and limiting productivity growth. This is perverse. It is a hugely inefficient and costly procedure for controlling inflation. It dooms the economy to slow economic growth, and often leads to recession. In many ways, the cure is worse than the disease.

At best, monetary restraint has only an indirect and slow-acting effect on inflation. Changes in interest rates by the Fed have very little direct influence on consumer spending. The principal effect is on borrowing by businesses. In any case, the time delay between changes in interest rates and changes in consumer prices can be many months. Success in fighting inflation comes only after months of delay during which economic growth has been slowed to the point where inflationary pressures are "wrung out" of market expectations. In the meantime, reduced investment and slow economic growth have a strong negative effect on unemployment and social welfare.

Controlling inflation through monetary restraint is fundamentally incompatible with the Peoples' Capitalism proposal to increase credit for investment by individual citizens. Unless we devise a better tool for controlling inflation, the massive increase in investment envisioned under Peoples' Capitalism would put the Federal Reserve at complete cross purposes with itself. To control inflation, the Fed would be making it more expensive for businesses to borrow for investment, while at the same time it would be making it easy for individuals to borrow for investment. This would simply shift investment spending from corporations to individuals, without increasing the overall investment rate. And, the effect on inflation would be minimal. Thus, we need a new tool for fighting inflation that does not involve restricting the supply of money for investment.

That brings us to the second feature of Peoples' Capitalism—savings.

Savings for controlling inflation

Consumer demand is the largest single component of demand and also potentially the most controllable. The most effective approach to controlling inflation would be to divert consumer income into savings. If the national savings rate could be temporarily increased to compensate for the short-term delay between the onset of investment spending and the social breakeven point, inflation could be controlled while investment spending is increased. One direct and straightforward approach would be to impose mandatory temporary savings withholding on consumer income. To achieve this goal, Peoples' Capitalism proposes a mandatory **Personal Savings Program (PSP)** to supplement monetary policy as a tool for controlling inflation.

The PSP would be designed to keep consumer demand aligned with the supply of goods and services in the market while the Federal Reserve is issuing credit for investment. Mandatory savings would restrain consumer purchasing power during periods of inflation by diverting some fraction of consumer income into personal savings in the form of Treasury bonds, certificates of deposit, or Individual Retirement Accounts. These saving instruments would earn market rates of interest, and would be redeemable without penalty after a period of at least five years. The effect would be to reduce demand and decrease inflationary pressures during periods when investment spending exceeds return on investment. The goal would be to reduce short-term inflationary pressure created by investment spending until after the social breakeven point.

The PSP savings rate would be indexed to the inflation rate. If inflation is higher than desired, savings would be automatically deducted from incomes in the amount needed to reduce consumer demand and bring inflation back to the acceptable rate. If inflation declines, the PSP savings rate would fall and consumer disposable income would immediately rise. The rate of savings would be automatically adjusted biweekly on the basis of the best available measure of inflation so as to minimize the time delay between measurement and action in the control loop that is the economic system. Modern computer technology and computer models of the economy make this technically feasible.

Mandatory savings is not a new idea. At the beginning of World War II, Keynes published a pamphlet entitled, *How to Pay for the War*.[67] In that small tract, he identified the "inflationary gap" created by resource constraints during the war effort, and promoted the device of "compulsory saving" and rationing to prevent price inflation. These proposals were largely adopted by the US Government in 1941. Keynes' 1940 publication is notable in that it provided the seeds of a theory of inflation to complement the "depression economics" of the *General Theory of Employment, Interest, and Money* published in 1936.[68] Similar to Keynes' proposal, the PSP would be designed control inflation by temporary withholding of savings from consumer income in the amount required to control inflation.

Thus, Peoples' Capitalism would give the Federal Reserve a powerful, fast-acting new tool for controlling inflation. This would free the Fed to use its current powers to influence interest rates for stimulating voluntary savings, attracting foreign investment, and preserving the value of the currency.

How much saving is required?

In the worst case, the rate of PSP savings would have to equal the rate of credit issued under the PIP minus the increased rate of productivity growth. In our example scenario, increasing the investment rate by 10 percent of GDP would increase the productivity growth by 3.8 percent, producing inflationary pressure of 6.2 percent. To reduce this to zero would require consumers to save an additional 6.2 percent of their income. This is a lot by US standards, but is quite small compared to the Chinese savings rate, which is in the neighborhood of 40 percent. This amount of saving would create

---

67   Keynes, John M. *How to Pay for the War: A radical plan for the Chancellor of the Exchequer.* London: Macmillan and Co., Ltd., 1940.

68   Keynes, John M. *General Theory of Employment, Interest, and Money,* Cambridge: Macmillan Cambridge University Press, 1936. Available online at http://ebooks.adelaide.edu.au/k/keynes/john_maynard/k44g/.

some hardship, but seems like an acceptable price to pay for full employment and an economy growing well over 6 percent per year. PSP savings would be graduated with income, so that the rich would have to save a larger percent of their income than the poor. And PSP savings would be indexed to inflation, so that if inflation were to remain under some acceptable level (say 3 percent), the PSP savings rate would be zero.

For a Near-Term Fix

Increasing the investment rate by 10 percent of GDP would certainly provide a near-term fix to the current economic crisis. $1.4 trillion of new money put directly into investment every year would pump liquidity into the financial markets. It would enable businesses to buy new equipment, build new production facilities, and hire new workers. Together with the PSP savings, it would boost real economic growth to 6 percent per year with a nominal inflation rate. It would create an immediate demand for labor that would rapidly reduce unemployment to a low level. Millions of additional workers would be needed for constructing new plants, building new equipment, developing new software, producing and selling new products, operating new businesses, and providing new services. And virtually all the jobs would be in the private sector. The only increase in government spending would be for policing the markets, prosecuting fraud, and funding research and development.

For Long-Term Results

Over the long run, the effect of compound interest magnifies the difference between different economic growth rates. Figure 4.3 compares the growth performance of the economy at growth rates of 3 percent, 6 percent, and 9 percent over a 50-year period.

An economy growing at the Congressional Budget Office estimated rate of 3 percent growth doubles in size in 23 years, and grows by 4.4 times after 50 years. An economy growing at 6 percent, doubles in size every 12 years, and is 18 times as big in 50 years. An economy growing at 9 percent doubles every 8 years, quadruples in 16 years, and grows by 74 times in 50 years. This is more than 17 times the CBO estimate.

Needless to say, these are vastly different results that would produce dramatically different effects on the economic, political, and military future of the United States. If the economy were to grow by 6 percent or 9 percent over the next five decades, all the doom-and-gloom predictions about the insolvency of Social Security, Medicare, and Medicaid would disappear. Unemployment would be no problem, and average incomes would soar. The defense budget could be increased, and taxes could be reduced without creating deficits. The debt could be paid down, and prosperity would reign.

Exponential Growth

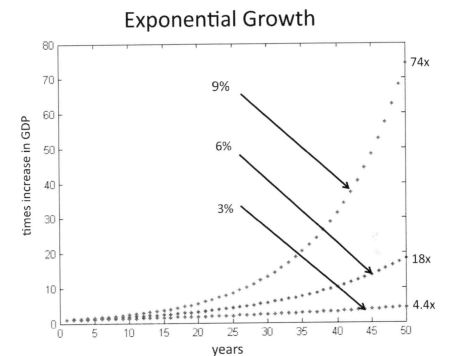

Figure 4.3. A plot of GDP growth at 3 percent,
6 percent, and 9 percent annual rates.

The current debate regarding how to balance the Federal Budget is being posed in terms of what percent of GDP should the government spend, and what percent of GDP should be collected in taxes. According to an editorial in the July 31 Washington Post by Matt Miller, a senior fellow at the Center for American Progress, the current rate of government spending is at a high of 24 percent, with taxes at a low of 15 percent (well below their long-term average of 18 percent).[69] Given current estimates of future growth, Miller predicts that government spending of about 28 percent of GDP will be required to maintain the current commitments of health care (Medicare and Medicaid) and Social Security to the American people as the population ages and the baby boomers retire en mass. However, this estimate is based on the current rate CBO estimates of 3 percent economic growth.

An economy growing at the CBO projected rate of 3 percent will grow only 1.8 times in twenty years. However, an economy growing at 6 percent will grow about 3.2 times in twenty years. The current GDP is $14.6 trillion.

---

69   mattino2@gmail.com.

If the economy grows by 3 percent per year, the GDP will be $26 trillion in 2030. However, if the economy grows at 6 percent, the GDP will be about $47 trillion in 2030.

Twenty-eight percent of $26 trillion is $7.2 trillion. But, $7.2 trillion is only 15 percent of $47 trillion. In other words, if the economy were to grow at 6 percent instead of 3 percent, we could maintain promised health care and Social Security commitments to a growing elderly population, maintain taxes at 15 percent of GDP, and still balance the budget. If the economy were to grow at 9 percent, we could increase health care and Social Security benefits, lower taxes, and still have a significant surplus.

For the younger age population, the more important effect would be income for individual investors. In 2008, GDP was $14.6 trillion. If we increase the investment rate by 10 percent of GDP, this translates into $1.46 trillion of new investment. Divided among 305 million citizens, this comes to $4800 to each citizen for investing in an approved mutual fund of their choice. Each year, that number would grow along with the economy at 6 percent per year.

The income that could be expected from this Personal Investment Plan by individual citizens is shown in figure 4.4. We assume a conservative 8 percent return on investment, a thirty-year payback period,[70] and a 2 percent surcharge on the outstanding balance for insurance and processing expenses by the banks. This leaves a net 6 percent return on the investment. Payback of the principal over 30 years would require 3.3 percent of the initial investment every year. This leaves a residual of 2.7 percent to be distributed as dividends to the individual investors.

As shown in figure 4.4, the residual income to the investor after loan repayments are deducted would be small at the beginning. After 10 years, it is about $1800 per year.[71] After 20 years, residual income grows to $5600 per year. After 30 years, residual income rises to $12,740 per year. At the end of 40 years, it rises to $27,870 per year. By the end of 50 years, the income floor for every citizen rises to $55,000 per person per year.

---

70   We should note that the 8 percent return on an investment that a shareholder in a mutual fund can expect is considerably below the 25 percent rate of return typically experienced by corporations on their internal investments. This is largely due to the fact that about two thirds of gross investment goes to prevent obsolescence. This is called "capital consumption" or depreciation. Also, the corporations typically do not pay out all their profits in dividends. In any case, 8 percent return on investment for individual investors is a reasonable assumption based on historical data.

71   First year = $128, second year = $267, third year = $217, fourth year = $580, fifth year = $755, sixth year = $945, seventh year = $1150, eight year = $1368, ninth year = $1600

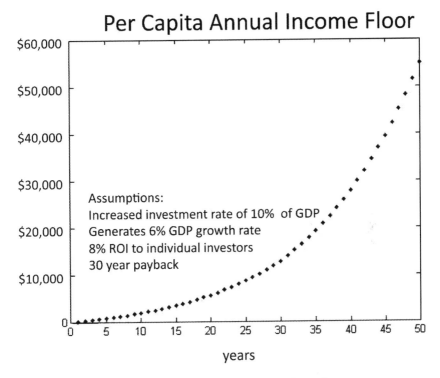

Figure 4.4. The per capita income floor for 10 percent of GDP increased investment.

Bear in mind that figure 4.4 shows only the income floor that would accrue to every citizen from return on capital investments financed by credit from the Federal Reserve. Most people would have additional income from wages, salaries, retirement benefits, or dividends, interest, capital gains, and rent from other investments over and above this floor. Because of the high rate of economic growth, jobs would be plentiful and well paid. The rising income floor would put money into the pockets of all consumers. Markets would expand and business profits would soar. It would be easy for those with energy and ambition to earn additional income, make additional investments, and become successful and wealthy. Everyone would benefit—rich and poor alike.

Of course, the effect on the poor would be the most profound. Within three decades, poverty would cease to exist. The poor would become bourgeois. The middle class would become rich, and the rich would become wealthy. The increased rate of investment would raise the market value of traditional stocks and bonds. Property values would rise. Markets would expand. New consumers with rising incomes would create new market opportunities. Income from investments would enable many to start their own businesses. Incentives for innovation and discovery would be great. Productivity growth would benefit everyone, because everyone would have a share of ownership of the means of production.

A Growing Portfolio of Capital Assets

Over the years, every citizen would acquire a significant portfolio of capital assets. Figure 4.5 shows the growth of a typical portfolio resulting from the PIP.

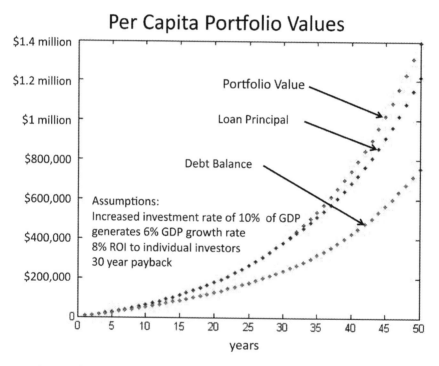

Figure 4.5. The per capita portfolio value, loan principal, and debt balance for 10 percent GDP increase in the investment rate.

At the end of the first year, portfolio value of individual investors would be just $4800. By the end of the first decade, portfolio value would rise to $63,000. At the end of the second decade, the portfolio value would be $176,000. After 30 years, portfolio value would be $379,000. After 40 years, the portfolio value would be $742,000. After 50 years, a typical portfolio value would be $1.4 million, with a debt balance of $757,000, producing an income stream of $55,000 per year.

The loan principal begins to fall below the portfolio value at the thirty-year point because the initial loans are being retired. The debt burden is the amount of investor loans that remain to be paid, and the difference between the portfolio value and the debt balance is the net equity which the individual owns. After fifty years, every individual would have a net equity of at least $636,000.

Increasing Investment by 20 percent of GDP

Assuming the smaller 10 percent increase worked out according to predictions, what would be the likely result if we increased the investment rate by 20 percent of GDP? This would double the amount of the loans to individuals for investment to $9600 for the first year, indexed by the GDP growth rate each year. Under this rate of investment, the rate of productivity growth and real economic growth could be expected to increase by another 3 percent to about 9 percent. The results are shown in Figure 4.6.

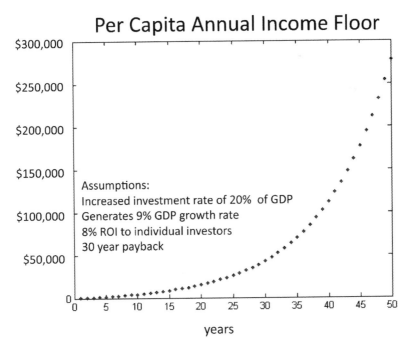

Figure 4.6. Predicted per capita income with increased investment of 20 percent of GDP.

After 10 years, the per capita income floor would grow to about $4200 for each individual. After 20 years, it would be $15,000 per person, or $60,000 for a family of four. This is three times the current poverty level in the United States. After 30 years, it would be $42,000. After 40 years it would be $112,000 and in 50 years, $277,000. Thus, with an increase in the investment rate of 20 percent of GDP, every child born in the United States could expect to have an annual income floor of more than $250,000 per year by their fiftieth birthday.

Portfolio value, loan principal, and debt balance for a 20 percent increase in the investment rate is shown in figure 4.7.

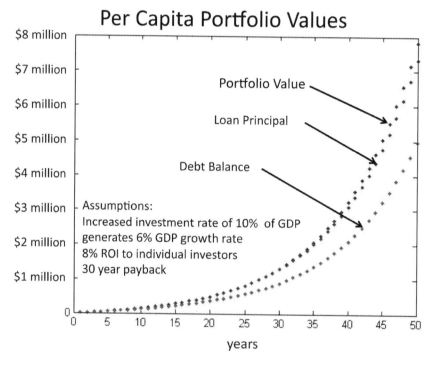

Figure 4.7 Per capita portfolio value, loan principal, and debt balance for a 20 percent increase in the investment rate.

Individual portfolio values would grow to $1 million in about 27 years, $2 million in 35 years, and reach almost $8 million in 50 years, with a net equity of $2.8 million after 50 years. PSP savings to prevent inflation would be on the order of 14 percent of income, worst case.

Benefits to the Economy

The benefits of Peoples' Capitalism to individual investors would be great. But as Mansfield and others have shown, the benefits to society as a whole would be much higher. Social benefits include the availability of higher-quality products at lower prices in the market, lower unemployment, higher incomes, lower deficits, lower tax rates, and more and better government services. Benefits to society also include a much stronger economy, a more affordable national defense, and a higher standard of living. For a society in recession, the biggest immediate benefit would be a dramatic reduction in unemployment.

In the short-term, infusion of credit into investment in the amount of 10 percent or 20 percent of GDP would provide more stimulus to the economy than any possible fiscal or monetary actions, such as tax cuts, unemployment benefits, grants to states for public-sector workers, or infrastructure construction and maintenance projects. Money invested directly in capital assets turns into orders for new machines, contracts for new plants, orders for more tools and materials, and jobs for workers. This in turn generates tax revenues to reduce government budget deficits.

The long-term benefits would be even greater. Long-term benefits include increased productivity from more modern and more efficient plants, processes, and equipment, growing streams of income to individual workers and their families, growing portfolios of capital assets for individual investors, and elimination of poverty. Long-term benefits also include energy independence from new sources of clean, renewable energy, modernized power distribution networks, high-speed rail and Internet services, better schools, increase tax revenue for national, state, and local governments, upgraded infrastructure, affordable health care, and solvency for Medicare, Medicaid, and Social Security.

In the long term, small farmers and fishermen who cannot compete with large agri-businesses could maintain their way of life because of a growing income stream from capital assets. Small business owners would be better able to survive recessions because of income from investments in the larger economy. Two-wage-earner families could revert to one-wage-earner families because of income from capital assets for everyone in the family. Children would receive income from birth that would contribute to family budgets when they are young, and finance their education as they enter young adulthood. By the time these children reach retirement age, they would be millionaires. Workers could retire early, or simply drop out of the labor force to go back to school, to start a business, or to pursue more desirable, but perhaps less remunerative, lines of work. Small towns could maintain vibrant economies that provide local opportunities for their youth because of income from investments in the outside economy. Opportunities would expand for pursuing dreams and creative ideas. Freedom from economic insecurity would enable the pursuit of happiness envisioned in the Preamble to the US Constitution.

Smoothing the boom-bust cycle

The investment mechanism proposed by Peoples' Capitalism would prove valuable in mitigating the boom-bust cycle that is endemic to the current capitalist system. One of the principal causes of instability in the current economic system is the volatility of private investors, whose investment decisions are biased by human emotions that fluctuate between fear and greed. If the central bank were to issue credit at a constant rate, it would significantly smooth out the fluctuations in the financial markets and stabilize the economy. There would still be fluctuations due to oscillations between fear and greed in the private market, but this would be reduced in magnitude by the steady flow of credit from the central bank through citizen investors. It may even be advantageous for the central bank to "lean against the wind" by issuing more credit for investment by individuals during periods when the growth rate falls below 6 percent, and cutting back during periods of "irrational exuberance."

PSP mandatory savings would also provide a powerful stabilizing influence. As soon as the economy begins to slow down, inflationary pressures will recede, and the rate of PSP savings would fall, giving consumers more discretionary income. When the economy heats up, inflationary pressures will rise, and the rate of PSP savings would increase, reducing consumer demand.

Because profits from technological innovation and productivity improvements would be distributed widely, public support for research and development of new innovations would be strong. Political support would build for government funding of research and development, both academic and industrial. Support for academic research is currently strong, but it could be usefully increased. Particularly if there were more support for transition of academic knowledge into practical uses in manufacturing, construction, agriculture, transportation, and energy.

Widespread ownership of capital assets would assure that everyone benefits from productivity growth and corporate profitability. Income from ownership is impervious to off-shoring of production. The owners receive dividends and capital gains regardless of where in the world production facilities are located. Thus, the benefits of free trade could be captured while the adverse impact from job losses in local industries would be mitigated by income from investments in profitable corporations.

Providing investment credit to individuals would also provide a new tool for dealing with economic crises. Instead of bailing out banks and brokerage houses through direct loans, which is politically unpopular, credit issued by the Fed could be routed through loans to individuals who could purchase stock in troubled financial institutions at distressed market prices. Then, when the crisis is over and the banks return to profitability, they would be partly owned by individual citizen investors who would share in the profits. This should be very popular.[72]

Thus, there are many social benefits to the Peoples' Capitalism policy of having the central bank issue credit to local banks for average citizens to invest in approved mutual funds. The benefits to the poor would obviously be the greatest. They would see a clear path out of poverty. The benefits to the middle class would be great. They would see a growing source of income in addition to their wages and salaries. The benefits to the rich would also be significant. The economy would grow rapidly. New consumers would enter the market, creating many new opportunities for business expansion. Dividend income to the poor would turn immediately into consumer demand for goods and services. The world's poor would become valued customers with cash to spend.

What would this better world be like?

In an economy based on Peoples' Capitalism, private ownership would remain the bedrock of the economic system. Capitalism would continue as the engine of prosperity. The free market would continue to allocate resources efficiently. Many would be rich, and opportunities for acquisition of great wealth would abound. More than half of the capital assets would be owned as they are now, by wealthy individuals, large banks, hedge funds, and big corporations. There would be many opportunities for those of greater-than-average talent, ambition, wealth, or power to become very rich.

What would be different is that average citizens would own a substantial percentage of the capital assets. Everyone would be able to accumulate a portfolio of capital assets large enough to provide a reliable and substantial source of income. After a few decades, ownership of capital would become the primary source of income for most people. There would be plenty of jobs

---

72   This type of transaction obviously would involve risk, as does any investment. The stock market price of any corporate entity should reflect the cost-risk-benefit of investing in that entity. The individual mutual funds on the approved list would make their own decisions regarding whether to buy or sell specific financial instruments available on the market.

available for those who want or need to work, but no one would be forced to work to feed their family. Children could anticipate that by the age of fifty they would have a net worth of more than $600,000 producing an income stream of more than $50,000 per year, over and above all other sources of income.

In the first half of the twenty-first century, rapid economic growth spurred by high rates of investment would provide strong demand for jobs. Rising incomes for the poor would provide the market demand to absorb the increased productive capacity. As income from capital assets grows, many families with medium to high incomes would revert to the single-breadwinner model. Many people would choose to drop out of the labor force, to go to school, to start a business, to pursue hobbies, or to perform volunteer work. This would leave plenty of attractive job opportunities for those who are ambitious and driven to succeed, or for those who like to work and want to make a positive contribution to society while being compensated for their effort. Today, many rich people work, not because they need the money, but because they want to make a difference and gain the power and prestige that comes from accomplishments—in business, government, law, education, medicine, sports, arts, and science.

By the second half of the twenty-first century, every individual in the United States would become economically secure and financially independent. This would have many benefits in terms of civil rights for minorities and women, democratic institutions, and individual liberties that derive from ownership of property. Under Peoples' Capitalism, no one would experience what Chapman calls the Extortion of "work or die."[73] Work would be by choice, not necessity. Rapid economic growth would create a strong demand for workers. But income from ownership of the means of production would provide an easy escape from less than attractive jobs. Jobs would be highly paid, and there would be strong incentive for automation. This would create rapid productivity growth and reduce opposition by organized labor to laborsaving innovations. Productivity growth would be viewed primarily as a means to generate higher income for worker-owners, not merely to reduce prices and improve quality of goods and services.

Under Peoples' Capitalism, economic growth would be strong and steady. The central bank would focus on promoting economic growth. The investment rate would be high and productivity growth rapid. Inflation would be controlled by mandatory PSP savings indexed to inflation. The cost of production would fall, quality would increase, profits would soar, and income

73   S. H. Chapman, 2005, *The Extortion: A Survival Manual for Hypersensitives in the 21st Century,* iUniverse, Lincoln, NE.

would flow to the owners who are the consumers that create demand for the goods and services that are produced.

In the next chapter, I will examine the progress in science and technology that have brought us to our current standard of living, and discuss the trends that will dramatically increase productivity and enable the rate of economic growth needed to make every citizen a capitalist.

CHAPTER 5 |||||||||||||||||||||||||||||||➡

The Science and Technology Enablers

S CIENCE IS KNOWLEDGE OF WHAT things are and how they work. Technology is knowledge of how to do things, how to make things, and how to make things do what we want them to. Science and technology are all that separates modern living from life in the stone age. Biologically, modern humans are not significantly different from humans of 20,000 years ago. Our bodies are somewhat larger today, but the computational mechanisms within our brains that determine our intellectual capacities have not changed much. It is not improved biology or advanced brain circuitry that makes our lives different from prehistoric humans. It is our superior science and technology. Today we live better because we know more, and because we have used that knowledge to improve productivity in the production of goods and services that people want and need.

Because of science and technology, average people today live in luxury that surpasses that of kings and popes a few centuries ago. We have indoor flush toilets, hot and cold running water, clear glass windows, cotton sheets and underwear, electric lights, central heating and air conditioning, microwave ovens, and automobiles with stereo sound systems, heated seats, and GPS navigation systems. We have airplanes that enable us to cross continents and oceans in a few hours. We have high-definition television, cell phones, and the Internet. We have antibiotics for disease and anesthetic for surgery.

Today we have personal computers with Internet browsers, cell phones, iPods, and iPads. Twenty years ago there was no Internet. Sixty years ago there were no personal computers. Ninety years ago there were no consumer electronics. One hundred and twenty years ago there were no electric lights.

Two hundred years ago there were no vehicles other than those propelled by wind or muscle power.

How did this happen? What is the secret? Why did the human race suddenly make more technological progress over the last 200 years than over the previous 200,000 years? What is the magic formula that has given our generation luxuries that have never before been enjoyed, even by the rich and powerful? The answer lies in the nature of knowledge. Knowledge is difficult to discover, but easy to copy. Knowledge does not wear out with use. When knowledge is shared with another, the knowledge of the giver is not diminished. Most importantly, knowledge is not subject to the law of diminishing returns; in fact, just the opposite. The more that is known, the easier it is to discover new things, and there is no limit to what there is to know. Knowledge grows exponentially. Each discovery builds on past discoveries, and enables future discoveries. Thus, the rate of discovery accelerates with each new piece of information.

Technological progress is on an exponential curve that is accelerating with each passing year.

More than half the scientists and engineers that have ever lived are alive today, and working with more and better tools and greater knowledge than ever before. Progress is rapid in so many fields that it is impossible for any one person, or even any single team of scientists, to keep track of all the new discoveries that are being made. Hundreds of journals and conference proceedings are publishing thousands of technical papers every month. Even the number of fields of science and technology is growing exponentially.

What is possible today surpasses what was possible in the past. What will be possible in the future will surpass what is possible today. The pace of technological progress is extremely rapid today, and will grow even more rapidly in the future. We live at a unique point in history. Within the next half century, we will have the technological and economic capability to eliminate poverty and create a world where everyone is physically and financially comfortable—but only if we recognize the potential inherent in the exponential growth of technology, and enact economic policies that exploit it.

The Long Journey

To fully appreciate how blessed our generation is over all those that have gone before, we need to understand something of the long and tortuous path that has brought us to this point in time.

Stone Tools

The rate of technological progress was very slow at the start. When little is known, it is difficult to discover new things. That is the nature of exponential growth. The first steps on the long chain of technological advances began with the development of stone tools by Homo habilis about 2.6 million years ago.[74] Before then, life must have been incredibly brutish. Agriculture did not exist. Fruit, root vegetables, and mushrooms could only be found growing wild. Game had to be captured by hand, and consisted mostly of insects, mollusks, and small animals that could be killed with a club. The technology of stone tools made it possible to kill larger game. The discovery that flint could be fashioned into sharp edges for spears and knives for killing and skinning large animals brought about a big improvement in the diet of early hominoids.

Fire

After the first stone tools, around two million years passed before the next big step occurred along the road to modern life—the domestication of fire. Without fire for cooking, all food had to be eaten raw. Without fire as a source of light, the nights were long and dark, except for moonlight and starlight. Without fire as a source of heat, many humans could not survive cold winters. Evidence for the controlled use of fire by Homo erectus dates back to about 400,000 BC give or take 100,000 years.[75] Fire enabled people to cook their food, warm their living quarters, and see at night. The domestication of fire began with naturally occurring fires from lightning strikes or volcanic activity. In those early days, fire had to be carefully maintained, and carried from place to place, because if the fire went out, it could not be restarted. The discovery of methods for starting fires, by rubbing pieces of wood together until the heat of friction caused shavings to burst into flame, or by striking a stone containing iron to produce sparks to light dry tinder, came much later, and represented a major technological advance. It made fire much easier to acquire and use.

Once the technology of starting fires became widely known, fire could be reliably restarted anywhere. It did not have to be carried from one encampment to another. Human groups could more easily move about in pursuit of food and water. Throughout human existence, technological advances in the making of tools and the use of fire have provided societies who mastered

---

74   Toth, Nicholas and Kathy Schick. "Overview of Paleolithic Archeology," In *Handbook of Paleoanthropology Part 3*, edited by Winfried Henke and Ian Tattersall, 1943-1963 Berlin: SpringerLink, 2007, Volume 3, Chapter 21.

75   Clark, J.D. and W. K. Harris. "Fire and Its Roles in Early Hominid Lifeways," *The African Archaeological Review,* 3(1985): 3-27

them a big advantage in the struggle for survival against the elements, and in the perennial competition with other humans over food, territory, and mating privileges.

Language

Language is critical for the development of technology. Spoken language enables knowledge discovered by one person to be passed on to others, but only face to face. Written language enables knowledge to be passed from one individual to another over great distances and long intervals of time. Written language is fundamental to the development of science and technology.

The first evidence of written communication appeared as paintings on the walls of caves 25,000 to 30,000 BC. These paintings appear to tell stories of events, and may have been used in religious rituals, or instructional material in preparation for the hunt. The earliest forms of writing used pictographs, where each symbol stood for an object or event.[76] The pictographic language of Egyptian hieroglyphs appeared around 4,000 BC. Pictographic languages persist today in traditional Chinese and Japanese kanji characters. The problem with pictographic languages is that the number of characters grows with the number of concepts to be expressed.

Alphabetic writing is based on spoken language where each letter corresponds to a phonetic sound. The first alphabetic writing appeared around 2,700 BC. The Phoenician alphabet appeared around 1,000 BC and spread across the Mediterranean where it evolved into the Aramaic, Hebrew, Greek, and eventually Latin alphabets.[77]

Agriculture

After the domestication of fire, more than 300,000 years passed before the development of agriculture. Until humans learned to plant seeds, tend crops, harvest produce, and domesticate animals, people lived as hunter-gatherers in small nomadic groups. Some followed the migration of animal herds. Some merely moved from place to place as the local supply of naturally occurring food became scarce.

It was about 10,000 BC that the transition from hunter-gatherer groups to settlements and agriculture began taking place in the Middle East.[78] Over the next several thousand years, hunter-gathers gradually began to

---

76   Daniels, Peter T. and W. Bright (eds.). *The World's Writing Systems*, New York: Oxford University Press, Inc., 1996.

77   Ibid.

78   Sauer, Carl O. *Agricultural Origins and Dispersals*, Cambridge, MA :MIT Press, 1952.

settle in villages and towns, cultivate crops, and develop irrigation systems. This provided the basis for crop specialization and food storage. With the development of fixed settlements, the rate of technology development began to pick up speed. The technology of planting and harvesting crops and the domestication of animals such as dogs, sheep, goats, pigs, and cows produced a significant increase in the quality and quantity of the food supply. This made it possible for large numbers of humans to gather in communities, towns, and cities. High population density communities gave rise to labor diversification, centralized administration, political structures, trade, art, architecture, culture, leisure, and writing. The first full-blown manifestation of these characteristics appeared in Sumerian cities in what is now southern Iraq around 3,500 BC.[79] Cities began to accumulate wealth, develop political and commercial enterprises, and be fortified with walls to protect against pillaging by roving groups of bandits and armies from neighboring cities. With the development of cities, the slope of the technology growth curve began to bend upward.

Metal Tools and Weapons

The first copper tools appeared shortly after the first permanent settlements around 9,000 BC. Almost six thousand years later, it was discovered that the addition of tin to melted copper produces bronze. Bronze is much harder than pure copper, and can take and maintain a much sharper cutting edge. Swords, spears, and battle axes made of bronze were much more deadly weapons than those made of copper or stone.

Iron melts at a much higher temperature than copper or bronze, and requires a furnace. So it was not until around 1,200 BC that the iron age began.[80] Even then the furnaces used were not hot enough to actually melt iron ore, but simply heated it to a point where particles of iron fell to the bottom of the furnace and formed a porous metallic substance that could be reheated and hammered to form a wrought iron object. The art of converting iron into steel is a complicated process with many variables. The ratio of iron ore to charcoal introduced into the furnace determined the characteristics of the product. It took another eight hundred years to discover how to build

79   Hallo, William and William Simpson. *The Ancient Near East*, New York: Harcourt, Brace, Jovanovich, 1971.

80   Waldbaum, Jane C. "From Bronze to Iron: The Transition from the Bronze Age to the Iron Age in the Eastern Mediterranean," *In Studies in Mediterranean Archaeology*, *LIV*. Göteburg: Paul Astöms Förlag, 1978.

furnaces that were hot enough to actually melt iron ore and control the impurities in the molten metal so as to reliably produce steel.

Steel is much stronger and more flexible than iron. It can bend without breaking and can take a much sharper and more durable edge. Mastery of the art of making steel led to many advances in weapons, and gave great advantage to the armies that possessed them. Steel weapons used by the armies of Rome were manufactured in what is now Austria. The Chinese began making steel about 400 BC, and steel was produced in India by 300 BC.[81]

In the ninth century AD, gunpowder was invented in China. By the thirteen century the technology of gunpowder spread from China to the Islamic world and to Europe. With the invention of gunpowder, bronze and steel were used for the manufacture of small arms and cannons.[82]

The use of steel for making objects other than weapons did not become common until efficient production methods were developed in the seventeenth century AD. The discovery that coke could be used instead of charcoal in the production of iron and steel allowed coal to be used for a fuel instead of wood for the smelting of pig iron. The invention of the blast furnace by Henry Bessmer in 1855 AD made steel inexpensive and plentiful. Today, steel is one of the most common materials in the world with more than 1,200 million tons produced annually. It is a major component in buildings, roads, bridges, tools, ships, railroads, automobiles, machines, appliances, and of course, weapons.[83]

As with other technologies, the manufacture of metal objects began slowly with very primitive methods and evolved exponentially over many centuries. At the beginning, discoveries took hundreds of years of trial and error. Today, advances in metallurgy and advanced materials occur every few months. The difference is that present-day metallurgists are guided by a vast, accumulated store of scientific and technical knowledge about the atomic structure and properties of metals. Knowledge of crystalline structure and bonding energies enable the development of new and improved materials. And the store of knowledge grows larger every year with nanotechnology and the ability to manipulate individual atoms.

81   Waldbaum, Jane C. "From Bronze to Iron: The Transition from the Bronze Age to the Iron Age in the Eastern Mediterranean," *In Studies in Mediterranean Archaeology, LIV.* Göteburg: Paul Aström's Förlag, 1978.

82   Jiasheng, Feng. *The Invention of Gunpowder and Its Spread to the West.* Shanghai: Shanghai People's Press, 1954.

83   *Steel Statistical Yearbook,* 2010, World Steel Association, Brussels: World Steel Committee on Economic Studies, 2010.

The Wheel

The invention of the wheel was one of the more significant events in the history of technology. The wheel was initially developed around 3,500 BC by the makers of pots in the early Sumerian cities.[84] Pots were initially made from coils of clay that were smoothed into the walls of the pots by squeezing. The ability to rotate the base on which the coils were formed increased productivity and quality in the making of pots. The use of an axle to keep the axis of rotation fixed gave rise to the potter's wheel. This enabled pottery to be fashioned more quickly with much thinner and more consistent walls and smoother shapes.

The first known use of the wheel for transportation occurred in the same culture about three hundred years later. The first use of wheels with spokes occurred in Egyptian chariots around 2,000 BC. The wheel first appeared in Europe around 1,400 BC.[85] The technology of the wheel and axel produced a major advance in the ability to transport goods. Over the next thousand years, the development of the yolk for oxen, and later harness for horses, made possible the use of animals to pull wagons and carriages. In Europe and Asia, wheeled carts, wagons, and carriages became ubiquitous. Wheels are not efficient without roads that provide a smooth surface for rolling. The Romans were master road builders. This facilitated the rapid movement of armies and the transportation of goods by wheeled vehicles.

Interestingly, the wheel was never invented by the great Inca, Aztec, or Mayan civilizations, and there is no evidence that the wheel existed anywhere in the Western Hemisphere before the Europeans arrived in the sixteenth century AD. This is a classic example of how knowledge is difficult to discover, but once known can easily be duplicated and exploited.

During the Industrial Revolution the wheel became a central component, not just for wheeled vehicles, but in machinery with rotating parts, such as gears, pulleys, cranks, pumps, drills, motors, and turbines. Today the wheel is indispensible to industry, commerce, and everyday life.

The Industrial Revolution

The invention of the steam engine enabled the conversion of heat into mechanical work. This was the key technology that triggered the Industrial Revolution. The steam engine was invented by Thomas Newcomen in 1712,

---

84   Yenne, B. *100 Inventions that Shaped World History*, San Mateo, CA: Bluewood Books, 1993.

85   Anthony, David. *The Horse, the Wheel, and Language: How Bronze-Age Riders from the Eurasian Steppes Shaped the Modern World*, Princeton, NJ: Princeton University Press, 2007.

and reduced to a practical device by James Watt in 1775. The steam engine allowed production processes requiring mechanical power to be located in places other than where falling water was available. Steam power was initially used to power reciprocating pumps for draining mines, but in the 1780s the conversion of reciprocating motion into rotary motion enabled steam engines to power factory machinery, ships, and railroad locomotives. [86]

The development of the steam engine created a need for machine tools that could cut metal parts more efficiently and with greater precision than could be achieved with hand tools. Machine tools were a critical technology that enabled the manufacture of tight-fitting cylinders and pistons for steam engines, and later for internal combustion engines.

In the latter part of the nineteenth century, steam engines were used to turn generators to produce electricity to power electric lights and electric motors for factory machinery and public transportation systems.

Other key inventions that contributed to the Industrial Revolution were the cotton gin, the flying shuttle loom, the Spinning Jenny, advances in the production of iron and steel, the McCormick reaper, the steel plow, the discovery of electricity, the railroads, and the internal combustion engine.

The rapid expansion of laborsaving technologies for spinning and weaving enabled a dramatic increase in the availability of cloth goods for average citizens at affordable prices, but it created a backlash among workers that had previously performed these tasks by hand. This resulted in the Luddite riots in England in 1811 and 1812, when mills were vandalized and pieces of factory machinery were burned by handloom weavers. The Luddite movement was brutally put down by the British Army in 1812 and many of the rioters were executed or deported to penal colonies.[87]

The railroad

It was 1825 when the first public steam railway in the world—the Stockton and Darlington Railway—opened for business in north east England. Five years later, George Stephenson and his son Robert built a steam locomotive called the *Rocket* that won a contest for the fastest locomotive on the Manchester–Liverpool line. The Stephensons subsequently became the preeminent builders of steam locomotives used on railways in the United Kingdom, the United States, and much of Europe. By 1850, railways connected all the major cities in Britain.[88]

---

86   Kirby, R., S. Withington, A. B. Darling, and F. Kilgour, 1990, *Engineering History*, New York: Dover Courier Dover Publications, 1990.

87   Bailey, Brian J. *The Luddite Rebellion*. New York: New York University Press, 1998.

88   Headrick, Daniel R. *Technology: A World History*, New York: Oxford University Press, 2009.

In 1830, the Baltimore and Ohio became the first railroad to evolve from a single line to a network in the United States.[89] The first transcontinental railway was completed in 1869.[90] Rail transportation made travel much faster and more convenient for passengers, and made hauling freight much faster and less expensive for industries and farms. The railroads were a key factor in the development of the western United States, and in the overall economic growth of the country. After World War II, improvements in internal combustion engines made diesel locomotives cheaper, more efficient, and more powerful than steam locomotives. This led railway companies to convert from steam to diesel power.

Modern high-speed passenger trains are powered by electricity. The Japanese Shinkansen "Bullet Train" introduced in 1964 was the first to exceed one hundred miles per hour. High-speed rail travel is now available in Japan, France, Spain, Germany, China, Italy, South Korea, and Taiwan—but so far, not in the United States.[91] In 2007, the French TGV set a speed record of 357 mph.[92]

The Internal Combustion Engine

The internal combustion engine is more fuel efficient and has a much greater power-to-weight ratio than the steam engine. This provides a more convenient power package for automobiles, trucks, trains, airplanes, and ships. Internal combustion engines come in three basic designs: piston engines, gas turbines, and rockets. Piston engines are primarily used for automotive, rail, and commercial ships. Piston engines for airplanes have largely been replaced by gas turbines. Rockets are primarily used for military missiles and to propel vehicles into outer space.

The most widespread application of the internal combustion engine is the automobile. The first versions of the automobile were literally horseless carriages. The earliest models were powered by steam engines, but these were quickly overtaken by internal combustion engines. The first patent for a gasoline-powered car was issued in 1885 for the Benz Motorwagon.[93] Three years later it went into production in Germany. The model T Ford began

---

89   Gordon, S.H. *Passage to Union — How the Railroads Transformed American Life, 1829-1929,* Chicago: Ivan R. Dee Inc., 1996.

90   McCague, James. *Moguls and Iron Men — The Story of the First Transcontinental Railroad.* New York: Harper & Row, 1964.

91   Hood, Christopher P. *Shinkansen – From Bullet Train to Symbol of Modern Japan.* London, Routledge, 2006.

92   "French High-Speed TGV Breaks World Conventional Rail-Speed Record," *Daily Le Parisien,* Feb 14, 2007.

93   Patent 37435, by Karl Benz for his 1885 Motorwagon.

rolling off the assembly line in 1908 in Detroit, Michigan.[94] Throughout the twentieth century the automobile has gone through a series of transformations that have culminated in the sleek, powerful, hybrid-powered, heated and air-conditioned cars of today that have numerous embedded computers, anti-lock braking systems, global positioning systems, computer route planning, satellite radio, stereo sound, and automatic cruise control. In the near future, cars and trucks will have computer-controlled collision avoidance systems, and soon after that, robotic chauffeur options.[95]

Electricity

Electricity is a critical enabling technology for modern society and for the Path to a Better World. Electricity provides light at night, heat in the winter, air conditioning in the summer, and refrigeration. It powers motors for subways, high-speed trains, hybrid vehicles, fans, pumps, and machine tools. It powers our computers, cell phones, cameras, radios, television sets, and toys.

Understanding of electricity developed over just the last 260 years. Static electricity produced by friction had been known since antiquity, but it was not until 1752 that Benjamin Franklin established the equivalence of lightning and static electricity.[96] In 1791, Luigi Galvani discovered that electricity was the medium by which nerve cells passed signals to the muscles.[97] In 1800, Allessandro Volta invented a battery made from alternating layers of zinc and copper.

In 1831, Michael Faraday began a series of fundamental experiments on electromagnetic induction, and in 1873 James Maxwell formalized the laws of electromagnetism in a set of mathematical equations that quantified the relationship between electricity and magnetism, predicted radio waves, and showed that light was an electromagnetic phenomenon.

In 1888, Heinrich Hertz validated Maxwell's equations experimentally by demonstrating the transmission and reception of electromagnetic waves in the laboratory. In 1896, Guglielmo Marconi invented the wireless telegraph. The key invention was the spark-gap transmitter, which simply transmitted long

94   Clymer, Floyd. *Treasury of Early American Automobiles, 1877-1925.* New York: McGraw-Hill, 1950.

95   Krisher,T. *GM Envisions Driverless Cars on Horizon.* Detroit: Associated Press, January 7, 2008.

96   Srodes, James. *Franklin: The Essential Founding Father.* Washington, DC: Regnery Publishing, 2002.

97   Baruzzi, A. C. Franzini, E. Lugaresi, and P. L. Parmeggiani, *"From Luigi Galvani to Contemporary Neurobiology: Contributions to the Celebration of the IX Centenary of the University of Bologna,"* Bologna, 27–28 September, 1988 (FIDIA Research Series) New York: Springer, 1990.

and short bursts of static that could be received and interpreted as telegraph signal dashes and dots. Marconi equipped ships with lifesaving wireless communications and established the first transatlantic telegraph service.

In 1879, Thomas Edison invented the electric light bulb. A year later, he formed the Edison Illuminating Company. Edison also invented the carbon microphone, the telegraph, the stock ticker, the phonograph, and motion pictures. A number of researchers including Edison performed experiments with light bulbs and discovered that the glowing filament in a vacuum emits electrons. In 1904, John Fleming discovered that a cold electrode placed into the same bulb would act as a diode, that is, a device that passes electric current in only one direction.

In 1907 Lee De Forest placed a third electrode known as a "grid" between the hot filament and the cold electrode. A voltage on this grid could control the amount of current flowing between the hot and cold electrodes. The resulting three-electrode device called a "triode" could be used to amplify voltages. With the proper feedback, a triode could also be used to generate radio frequency oscillations that could be modulated with acoustic frequencies. Thus, triodes could enable both the transmission and reception of radio signals carrying voice communications and music.[98]

Improvements in the design of vacuum tubes enabled more and more sophisticated transmitters and receivers capable of generating very high frequencies and switching on and off extremely fast. These advances led to the development of commercial radio, television, radar, cell phones, cable and satellite communication system, and electronic computers.

The Computer

The computer may turn out to be the most important technological development of all time, because it enables the substitution of computational power for brain power in the control of machines and processes. The theoretical foundations of mechanical computation were laid down by mathematicians such as Descartes, Pascal, Boole, Whitehead, and Russell, who developed formal systems of logic and algorithmic procedures.

The first commercial machine controlled by a program was the automatic loom invented by Joseph Jacquard in 1801.[99] It was controlled by punched cards with rows of holes, where each row on the card corresponded to one row of the design in the cloth. Multiple rows of holes were punched on each

---

98   Kirby,R., S. Withington, A. B. Darling, and F. G. Kilgour, 1990, *Engineering in History*, New York: Courier Dover Publications.

99   Essinger, James. *Jacquard's Web: How a Hand-Loom Led to the Birth of the Information Age*. Oxford: Oxford University Press, 2004.

card, and many cards were strung together in a chain. The resulting sequence of punched holes was a program that controlled the machine to produce the desired pattern.

Charles Babbage is widely considered the "father of the computer." In 1822 he designed what he called the "difference engine" to compute values of polynomial functions. He later designed an "analytic engine" that was to operate with a program provided by punched cards. His designs were for mechanical devices that were too complex for the technology available at the time, and none of his machines were actually completed during his lifetime.[100]

Calculating machines came into widespread use after William Burroughs received a patent for his adding machine in 1888 and founded a company that later became the Burroughs Corporation.[101] In 1889, Herman Hollerith developed a mechanical tabulator based on punched cards and electrical card readers. It was used by the US Census Office to tabulate statistics from millions of pieces of data collected during the 1890 census. Hollerith later founded a company that became IBM.[102]

Machines for encrypting and decoding messages were widely used on both sides during World War II. The first digital computers were developed during that conflict for breaking encrypted messages and for computing ballistic trajectories for artillery. Alan Turing was one of the mathematicians that developed techniques to break the codes used by the German Enigma encryption machines.[103] These early computers used electric relays for switches and hence were very slow compared with modern machines.

Vacuum tube computers developed during the late 1940s and 1950s were much faster than the machines based on mechanical relays. In 1945, von Neumann, Eckert, and Mauchly developed the concept of stored program computers, where programs could be stored in the same type of memory as data.[104] This made possible computers that could be reprogrammed to perform different functions.

During the 1950s, vacuum tubes were replaced by newly invented transistors, initially to increase the reliability of intercontinental missile

---

100   Hyman, A. *Charles Babbage, Pioneer of the Computer*, Princeton: Princeton University Press, 1982.

101   Campbell-Kelly, Martin and William Aspray, *Computer: A History of the Information Machine*, New York: Westview Press, 2004.

102   Heide, L. *Punched-Card Systems and the Early Information Explosion, 1880–1945*. Baltimore, MD: Johns Hopkins Press, 2009.

103   Copeland, B. L. *Colossus: The Secrets of Bletchley Park's Codebreaking Computers*, Oxford: Oxford University Press, 2006.

104   Rojas Raul and Ulf Hashagen, eds. *The First Computers: History and Architectures*, Cambridge, MA: MIT Press, 2000.

guidance systems. In the 1960s, the ability to put many transistors on a single piece of semiconductor material gave rise to integrated circuits. Eventually, this technology progressed to the point where an entire computer could be deposited on a single chip of silicon. In the 1970s, this gave rise to an entire family of microcomputers that enabled personal computers that are now ubiquitous in offices, homes, and classrooms. Today, multiple computers can be interconnected on a single chip.

At each step in this progression, the speed of computation and the amount of memory available increased exponentially. Between 1950 and 1990, the speed and memory capacity that could be procured for a given amount of money grew by more than a factor of ten every decade. Since then, the growth rate has accelerated to the point where computational power per unit cost doubles every eighteen months and grows by a factor of ten every five years. Although some physical barriers will prevent the current technology from maintaining this rate of growth for much more than another decade, new technologies of multi-core computers, hierarchical hardware architectures, and new materials should allow this growth rate to continue for at least two more decades. Beyond that, quantum computing and other exotic breakthroughs seem likely to allow the exponential growth in computational power to continue well into the future.

Computer applications

Computers have already had an enormous impact on productivity in the manufacturing industries. Computer-aided design systems have eliminated the need for mechanical drawings. Computer-aided engineering systems enable simulation of static and dynamic forces and stress in components and systems, and enable optimization of products. Computer-aided production planning and inventory control systems dramatically improve efficiency and productivity in the flow of goods and materials throughout the business enterprise. Computer control of machine tools, robots, and production lines provides optimal control of activities on the factory floor. Many of the productivity gains in the manufacturing industries over the last thirty years have been due to the introduction of computers into the manufacturing processes. This trend will certainly continue as computer algorithms become more sophisticated and able to make more sophisticated decisions. Built-in tests and diagnostic capabilities will improve reliability and reduce maintenance costs.

Major retail corporations have computer-controlled supplier networks whereby sales at the checkout counter automatically order replacement merchandise. Computer-controlled warehouses are already in commercial operation. In the future, computer-controlled robots will fill orders, schedule deliveries, load the trucks, and deliver packages to the customer's front door.

Computer technology has already replaced human brain-power in low-level control of industrial machinery. In most modern machines, computers and electronic circuits provide precision controls that far exceed the accuracy, speed, and concentration powers of human workers. Most chemical plants are controlled by thousands of sensors and electronic control systems that monitor pressure, temperature, and flow rates, and make hundreds of adjustments every second with accuracy and dependability that far exceed the capabilities of human operators. Humans still supervise these computerized systems to monitor safe operation and make high-level decisions based on production goals, company policy, and market prices, because computers still cannot be trusted to make these kinds of value judgments. However, research into mathematical models and computer algorithms that can duplicate, or even surpass, human performance in these high-level decision and control problems is an active field of research, and progress is substantial. As computers become more powerful and algorithms more sophisticated, it seems likely that computer systems will eventually be capable of even the kinds of situation assessment, decision-making, planning, and control that currently require human judgment.

Computers are embedded in household appliances, cars, trucks, and airplanes. They are used in the control of chemical plants, power plants, and electrical power grids. The impact of advanced computer software is being felt everywhere in offices and businesses. The personal computer has significantly improved productivity in secretarial work, accounting, and purchasing. Telephone switchboard operators no longer exist. Travel agents and ticket clerks have been replaced by online reservations systems and electronic tickets. Productivity in communication services is evident in the explosive growth of the Internet and the cell phone industry. Personal and business communications have revolutionized the way people interact with each other. Meetings can be held without traveling long distances. Telecommuting is becoming popular as a way of working that eliminates the need to physically travel to and from the workplace.

Computers are ubiquitous in the financial services industry. They are used for money transfers, stock market transactions, credit card purchases, and checking accounts. Cell phones, Blackberrys, and iPads all contain powerful embedded computers. Millions of computers are involved in cell phone routing and moving information over the Internet. The emerging technology of "cloud computing" will enable the linking of thousands of computers for solving really big scientific and engineering problems.

In the construction trades, computer-aided design systems are used by architects and structural design engineers for optimizing the design of roads, bridges, skyscrapers, and homes. Construction companies routinely use computers for planning and scheduling construction site operations. Intelligent machines and tools with embedded computers will be used for surveying, measuring, cutting, fitting, digging, grading, erection of structures, and finishing of interior and exterior surfaces. Within a decade, computer-controlled robots will begin to appear on the construction site. This will enable major productivity improvements in the construction of all types of structures. Computer-controlled robots will eventually be able to lay foundations, build walls, install electrical wiring, plumbing, heating, and air conditioning, set roof supports, and apply roofing materials more cost effectively than human workers. Advances in intelligent systems engineering will enable computer-controlled machines and automation systems to mine raw materials, manufacture products, construct roads and buildings, transport goods, and recycle waste.

On the farm, intelligent machines will be employed for plowing, planting, tending crops, caring for animals, harvesting, processing of food, shipping to markets, and delivery to consumers. Computer-controlled machines will be used in mines and drilling operations, especially in deep-sea drilling. Computers will control plants that process biomass into carbon neutral fuels.

Researchers in robotics, automation, and intelligent control systems have learned how to encode knowledge, skills, and abilities in computer data structures, and how to program intelligent systems to be capable of performing complex operations in dynamic, real-world, uncertain, and sometimes hostile, environments. Reference model architectures and software development methodologies and environments have evolved over the past three decades that provide a systematic approach to engineering intelligent systems. [105] [106] These advances will soon lead to computer-controlled automobiles and trucks that can operate safely on roads and highways. In 1994, an autonomous vehicle control system developed at the Military University in Munich enabled two autonomous vehicles to drive more than one thousand kilometers on a three-lane highway in normal heavy traffic at speeds up to 130 km/h (82

---

105   Albus, James S., et al. *4D/RCS Version 2.0: A Reference Model Architecture for Unmanned Vehicle Systems, NISTIR 6910*, National Institute of Standards and Technology, Gaithersburg, MD. Available online at http://www.james-albus.org/docs/4DRCS_ver2.pdf.

106   Albus, James S. and Anthony J. Barbera, "RCS: A Cognitive Architecture for Intelligent Multi-Agent Systems," *Annual Reviews in Control*, Vol. 29, Issue 1, (2005): 87–99. Available online at: http://www.sciencedirect.com/science/article/pii/S1367578805000027.

mph).[107] In 2001, the Army Research Lab Demo III experimental unmanned ground vehicle program demonstrated the ability of fully autonomous vehicles to drive through complex environments such as woods, dirt roads, brush, and dry streambeds at tactical speeds. In 2005, an autonomous SUV using software generated by Stanford University students was able to win a DARPA-sponsored competition by driving 132 miles on dirt roads over mountainous desert terrain in less than 7 hours.[108] Two years later, a vehicle fielded by the Carnegie Mellon University was able to drive 60 miles through an urban environment while negotiating with other vehicles, avoiding obstacles, and obeying traffic regulations.[109]

Within a decade, automobiles, trucks, trains, ships, and airplanes with computer control systems will be safer and more reliable than those with only human operators. Every year in the United States, over thirty thousand people are killed in traffic accidents.[110] This is roughly equivalent to a 9/11 terrorist attack every month. Worldwide, more than one million people are killed every year in traffic accidents and more than 50 million are injured.[111] Most of these accidents are caused by human error. Intelligent cars and trucks with radar and computer vision systems have the potential to reduce the number of traffic deaths and injuries by more than half. Within two decades, automobiles with a robot chauffer option may appear on the market. These vehicles will be capable of completely taking over the driving task when requested to do so. This capability will enable a person to simply tell the car where they want to go and let the computer do the driving. At the desired destination, the car will drop off the passenger, find a parking place, and wait to be called by cell phone for a return trip.

These technological advances will enable dramatic improvements in productivity that advance the ability of the economy to produce the goods and services that people want and need. They will enable the production of material wealth on a scale that has never before been possible.

---

107   Dickmanns, Ernst. *Dynamic Vision for Perception and Control of Motion*, London: Springer-Verlag, 2007.

108   *Journal of Field Robotics, Special Issue on DARPA Grand Challenge, Part 1*

109   *Journal of Field Robotics, Special Issue on DARPA Grand Challenge, Part 2*

110   National Highway Traffic Safety Administration Fatality Reporting System, http://www-fars.nhtsa.dot.gov/Main/index.aspx.

111   Mundel, E.J. "U.N. Seeks to Curb World's Traffic Deaths," *Washington Post Newspaper*, April 1, 2008. Article available online at http://www.washingtonpost.com/wp-dyn/content/article/2008/04/01/AR2008040101507.html.

Computer evolution

Intelligent behavior in machines is evolving very rapidly because it is driven by intelligent design. Human designers use the best available tools for designing each new generation of computers. Hardware and software design tools are constantly improved to increase their ability to explore the design space for the best materials, the optimal algorithms, and the most powerful computational mechanisms. New materials are being developed that enable new capabilities.

Automated machines are able to build automated machines that are more capable and efficient than themselves. This drives evolution in machine performance and leads to rapid growth in the number of intelligent machines. As each generation grows more capable and more widely used than the one before, growth in numbers and capabilities will be exponential.

A precursor to this phenomenon can already be observed in the computer industry where computers are used in the design, manufacture, testing, and marketing of computers. Exponential growth in the computational power of computer systems is a well-established fact that has been codified in Moore's Law[112] which empirically observes that computational power grows by a factor of 2x every eighteen months, and by an order of magnitude (10x) every five years. This is a phenomenal rate of growth. It produces a 10,000 times increase in computational power per unit cost every twenty years.

This rate of exponential growth has given rise to the concept of a "singularity," that is, the point in time when computers exceed the intellectual power of the human brain—and soon thereafter exceed the collective intelligence of the entire human race. Ray Kurzweil is the most prominent of the proponents of this idea. In his book, *The Singularity Is Near*, he lays out the case for the Singularity occurring before the halfway point in the twenty-first century.[113] Kurzweil extends Moore's Law to the exponential growth of technological progress in general. Relying almost entirely on empirical data, he argues persuasively that not only the return, but the *rate* of return on investment in technology is increasing exponentially.

In 2009, Kurzweil, together with Google and the NASA Ames Research Center established a Singularity University "to leverage accelerating technologies to address global issues."[114]

---

112   Hutcheson, D. G. "Moore's Law: The History and Economics of an Observation that Changed the World," *The Electrochemical Society INTERFACE*, Vol. 14, No. 1, Spring 2005, pp. 17–21.

113   Kurzweil, Ray. *The Singularity Is Near*. New York: Penguin Group, New York, http://www.singularity.com/.

114   Singularity University website http://singularityu.org/about/overview/.

Reverse Engineering the Human Brain

While it is not possible to predict all the future applications of computers, there is one development that will change everything. When it becomes possible to build computers with the perceptual, cognitive, and decision-making capabilities of the human brain, the world will become a different place.[115]

The raw computing power required to reverse engineer the human brain will soon be available. Estimates of the computational power of the human brain range from 10^{12} to 10^{16} operations per second (ops).[116, 117] Today's biggest supercomputers have already achieved 10^{15} ops. Thus, supercomputers are already well within the estimated range of computing power of the human brain. Computer Graphics Processing Units (GPUs) used for computer games operate at about 10^{13} ops, and general-purpose, laptop-class machines operate at about 10^{10} ops. If Moore's Law holds for another two decades, by 2030 supercomputers will be capable of 10^{19} ops, GPUs will operate at 10^{17} ops, and laptop-class machines will be in the neighborhood of 10^{14} ops.

Of course, more than raw computing power will be required for duplicating the capabilities of the human brain. How does the brain actually work? How can three pounds of gelatinous protein generate the incredible experience of consciousness? How are signals from millions of sensors in the eyes, ears, skin, nose, and tongue transformed into an internal representation that we perceive as external reality? How does the brain do this? Can we discover how to build machines and software that simulate what the brain does? If we can, it will change everything. The future will be very different from the past.

Reverse engineering the brain will require a deep understanding of how information is represented and how computation is performed in the brain. There are many unanswered questions. What are the functional operations? What are the data structures of knowledge? How are messages encoded? How are relationships established and broken? How are images processed? How does the brain transform signals into symbols? How does the brain understand human speech, and encode thoughts into language? How does the brain generate the incredibly complex, colorful, dynamic, internal representation that we consciously perceive as external reality? These are profound questions that must be answered before computers can rival the capabilities of the

---

115   Albus, James S. "Reverse Engineering the Brain," *International Journal of Machine Consciousness,* Vol. 2, No. 2, 2010, pp. 1–19.

116   Moravec, H. *Mind Children: The Future of Robot and Human Intelligence,* Cambridge, MA: Harvard University Press, 1998.

117   Albus, James S. and Alexander M. Meystel, *Engineering of Mind: An Introduction to the Science of Intelligent Systems,* New York: John Wiley & Sons, 2001.

human brain. But again, progress is rapid and knowledge of how the brain works is growing exponentially.

Neuroscience

Over the last century, researchers in neuroscience and cognitive psychology have made great strides in understanding the computational mechanisms of perception, cognition, reasoning, motivation, decision-making, planning, and control of behavior that take place in the brain.[118]

Neuroscience is currently one of the most active fields of scientific research, and progress is rapid. It was less than 150 years ago that Paul Broca and Carl Wernicke[119] discovered that different regions of the neocortex are responsible for quite specific different functions. Wernicke's area is responsible for speech understanding, and Broca's area is responsible for speech generation. It was 1890 when Ramon y Cajal[120] provided clear evidence that the neuron is the fundamental computational unit in the brain. It was 1903 when Ivan Pavlov first reported his work on the "conditioned reflex" in dogs.[121] During the 1930s and 1940s, B. F. Skinner studied "operant conditioning" on rats and pigeons,[122] and together with John Watson and Edward Thorndike founded the school of behaviorism.

In 1943, Warren McCulloch and Walter Pitts proved that neurons can compute all of the basic functions performed by digital computers.[123] In 1952, Alan Hodgkin and Andrew Huxley explained the ionic mechanisms underlying the initiation and propagation of action potentials in the neuron axon.[124] In the 1960s, David Hubel and Torsten Wiesel discovered a number

---

118   Albus, James S. "A Model of Computation and Representation in the Brain," *Information Sciences* 180 (2010), pp. 1519–1554, Available online at http://www.james-albus.org/docs/ModelofComputation.pdf.
       Albus, James S. "Toward a Computational Theory of Mind," *Journal of Mind Theory*, Vol.0, #1, 2008, pp. 1–38.

119   Kandel, Eric R., James H. Schwartz, and Thomas M. Jessell. *Essentials of Neuroscience and Behavior*. New York: McGraw-Hill, 1995.

120   Cajal, S.R. (1909–11) Translated by Neely Swanson and Larry W. Swanson. *Histology of the Nervous System of Man and Vertebrates* New York: Oxford University Press, 1995.

121   Pavlov, I.P. *Conditioned Reflexes: An Investigation of the Physiological Activity of the Cerebral Cortex*. Translated and edited by G. V. Anrep, London: Oxford University Press, 1927.

122   Skinner, B.F. *The Behavior of Organisms: An Experimental Analysis*. Cambridge, MA: B. F. Skinner Foundation, 1938.

123   McCulloch, Warren S. and Walter Pitts, "A Logical Calculus of the Ideas Immanent in Nervous Activity," *Bulletin of Mathematical Biophysics*, 5, Issue 4. New York: Springer (1943):115–133.

124   Hodgkin, A. and A. Huxley. "A Quantitative Description of Membrane Current

of the computational functions that the visual cortex performs on the images received from the retina. In 1991, Daniel Felleman and David Van Essen published a circuit diagram of the visual cortex of the macaque monkey showing that the visual cortex consists of 32 different processing arrays arranged in 12 hierarchical layers and connected by 305 nerve bundles.[125]

Recent work has found specific regions in the brain that recognize faces, and detect shapes such as hands.[126] Other regions can recognize behaviors performed by others.[127] Still other regions compute worth and assign emotional value to objects and events in situations and episodes.[128]

Each neuron in the central nervous system receives input from a specific set of other neurons, and sends information to another specific set of neurons. Connectivity of neural circuits determines functionality. Neurons are clustered into modules that perform specific computational functions. In the posterior brain, clusters of neurons are arranged in hierarchical arrays for processing spatial-temporal visual, tactile, and acoustic images. These neural processing arrays transform input from sensors into an internal representation that we perceive to be external reality.[129]

Simultaneously, similar hierarchies of arrays in the frontal neocortex use the brain's internal representation to make behavioral decisions, set goals and priorities, generate plans, and control behavior. Long-range plans are developed in the prefrontal cortex. Goals, tasks, and priorities generated by the prefrontal cortex are decomposed into behavioral commands through a behavior generating hierarchy consisting of the frontal, premotor, and primary motor cortices. From there, motor commands are sent to computational modules in the midbrain, and then to the spinal cord where final motor neurons drive muscles in the execution of behavior. At each echelon in this hierarchy, the limbic system evaluates the cost, risk, and benefit of planned actions. At each echelon behavioral commands are monitored by the cerebellum and other midbrain computing centers where they are compared with feedback from stretch, tension, and velocity sensors in the muscles and

---

    and Its Application to Conduction and Excitation in Nerve." *Journal of Physiology,* Vol. 117, (1952): 500–544.

125   Felleman, D. J. and D. C. Van Essen. "Distributed Hierarchical Processing in the Primate Cerebral Cortex." *Cerebral Cortex,* Vol. 1, (1991):1–47.

126   Koch, Christof. *The Quest for Consciousness: A Neurobiological Approach.* Engelwood, CA: Roberts & Company Publishers, 2004.

127   Rizzolatti, G. and L. Craighero, "The Mirror-Neuron System," *Annual Review of Neuroscience,* Vol. 27, (2004): 169–92.

128   Kandel, Eric R., James H. Schwartz, and Thomas M. Jessell. *Essentials of Neuroscience and Behavior,* New York: McGraw-Hill, 1995.

129   Albus, James S. "A Model of Computation and Representation in the Brain," *Information Sciences,* Vol. 180, (2010):1519–1554.

joints.[130] Output from all these computing centers controls the activity of the final motor neurons such that desired actions are actually accomplished despite unexpected perturbations from the environment. Each region of the brain performs a specific function in this overall process. The various regions are interconnected in a network of processors that work together to produce the phenomena of mind, that is, perception, cognition, self consciousness, memory, emotion, and volition.

Since the early 1990s, modern techniques of functional Magnetic Resonance Imagery (fMRI),[131] electroencephalograms (EEG),[132] and advanced neurophysiological and neuroantomical studies have been revealing details regarding the functionality and connectivity of various regions in the brain. fMRI has made it possible to measure the change in blood flow related to neural activity in various parts of the brain and spinal cord of humans and other animals. EEG signals enable scientists to measure temporal relationships between activities in different regions in the brain. These technologies have made it possible to visualize which parts of the brain are involved in which kinds of mental activity, and how these regions are interconnected. The technology of diffusion spectrum imaging developed since 2001, coupled with advanced neurophysiological and neuroantomical studies, promise to enable the construction of detailed maps of the major communications pathways in the brain.[133]

Of course, much about the brain is still unknown, and many of the secrets of the mind and brain will remain a mystery for decades, perhaps even centuries. But enough is known to begin serious scientific studies and engineering efforts to reverse engineer the human vision system. There are several research programs currently underway, and more being contemplated, in the United States, Europe, and the Far East. In the United States, the Defense Advanced Research Projects Agency (DARPA), the National Security Agency (NSA), the Office of Naval Research (ONR), the National Science Foundation (NSF), and many university, government, and private research laboratories are actively pursuing research that either directly addresses the scientific understanding of the brain, or supports it in a variety of ways. The

---

130   Albus, James S. "A Theory of Cerebellar Function," *Mathematical Biosciences*, Vol. 10, (1971): 25–61.

131   Raichle, Marcus E. and Mark A. Mintun. "Brain Work and Brain Imaging," *The Annual Review of Neuroscience*. Washington University Library (2006): 449–476.

132   Niedermeyer, E and F. L. da Silva. *Electroencephalography: Basic Principles, Clinical Applications, and Related Fields*, Baltimore, MD: Lippincot Williams & Wilkins, 2004.

133   Wiegell, W., T. Reese, D. Tuch, G. Sorensen, and V. Wedeen, "Diffusion Spectrum Imaging of Fiber White Matter Degeneration," Proceedings of International Society for Magnetic Resonance in Medicine, Vol. 9, (2001): 504.

US National Academy of Engineering has identified "Reverse Engineering the Human Brain" as one of its Grand Challenges.[134]

Given the growing body of knowledge about the brain, combined with understanding of the structure and function of the biological computational mechanisms, and the growing power and speed of modern computers, success in reverse engineering the brain seems plausible, even probable within two or three decades.

Economic Implications

Reverse engineering the brain will eventually enable inexpensive computers to perform the kinds of perceptual and intellectual tasks that only humans can perform today. This will have a profound impact on commerce, industry, and military power. Computers will be able to drive cars and trucks, fly airplanes, understand language, read and write reports, and perform cognitive tasks that today can only be done by highly trained professionals. Computer programs may become as skilled and competent as doctors, lawyers, managers, and executives. Computer-controlled weapons systems may grow as capable and effective as well-trained soldiers, sailors, pilots, and submariners. P. W. Singer in his book *Wired for War* describes how the robotic revolution will transform the art of war in the twenty-first century.[135]

Perhaps the most important impact of human-level artificial intelligence will be felt in the industrial and commercial world. As noted above, advanced automation has already had an enormous effect on productivity and quality in manufacturing industries. As robots become lighter, more mobile, and more flexible, with better sensing, perception, decision-making, and planning and control capabilities, they will be able to deal with the uncertainties and complexities of the real world. Robots with human-level intelligence will be able to work as assistants, to fetch and carry, to provide tools and fixtures, to perform simple tasks, and to supply measurement capabilities to human workers. As their reasoning and decision-making capabilities improve, they will be able to function as mid-level business managers, accountants, lawyers, engineers, software developers, and information technology specialists. They will be able to perform human-level tasks in agriculture, construction, mining, transportation, retail, health care, education, and service industries. This will not just improve the productivity of human workers. It will provide skills and abilities previously unique to human workers. The advent of truly intelligent

---

134   National Academy of Engineering list of engineering grand challenges, http://www. engineeringchallenges.org/.

135   Singer, P.W. *Wired for War: The Robotics Revolution and Conflict in the 21ˢᵗ Century*, New York: Penguin Press, 2009.

robots will effectively increase the size of the work force beyond that provided by human workers. It will be as if a new race of workers were to visit earth and work for nothing more than the cost of their manufacture and maintenance. The impact will be felt everywhere.

Once the cost of a computer-controlled machine is less than the cost of a human worker, and the job skills and productivity of the machine exceed those of the human worker, then the number of robot workers will rapidly increase. The increase will be exponential, and each generation of intelligent machines will be more powerful and less expensive than the one before. Computer-controlled machines will be able to learn new knowledge and skills by simply downloading software. Years of training will not be necessary. Once the knowledge, skills, and abilities of human workers are encoded in software, they can be loaded into a computer brain in a matter of minutes or hours, as opposed to months or years for humans.

The result will be that productivity will soar, and the need for human workers will decline. If an intelligent robot costing $100,000 can do the job of a human worker making $50,000 per year, there will be an overwhelming economic case to replace the human. The robot can work about 160 hours per week for 50 weeks a year for around 5 years. This is 40,000 hours, which works out to $2.50 per hour. Once technology advances to the point where a computer can duplicate the computational capabilities of the human brain, the relationship between labor and capital expressed in the classical economic theory will completely be turned on its head.

The potential economic impact of this technology is impossible to estimate. It almost certainly will increase total factor productivity at an unprecedented rate. As the cost of intelligent machines falls and their capabilities increase, a 1 percent GDP increase in investment in computer automation might produce a 1 percent increase in total factor productivity, or even greater. In that case, economic growth rates of 10 percent, 12 percent, or even 15 percent could easily be achieved. At a growth rate of 15 percent, the GDP doubles every 5 years, quadruples every 10, and rises by more than one thousand times in 50 years.

Unfortunately, these levels of productivity growth would be devastating to workers unless they have a substantial source of income other than wages and salaries. If intelligent robots can do the work of skilled human workers, from mid-level managers to factory floor workers, for $2.50 per hour, it would create catastrophic unemployment problems—and not over a period of centuries, but over a period of decades.

Exponential Nature of Knowledge

The moral of the story of technology development is that progress is exponential. In the beginning, technology development was extremely slow. Two million years passed between the invention of stone tools and the domestication of fire. Another 400 thousand years passed before the development of agriculture or written language. But once humans began living in cities, the rate of discovery accelerated. The ancient civilizations of Egypt, China, and Persia made advances in technology every hundred years or so. The Greeks and Romans averaged significant improvements in technology every several decades. With the advent of the Renaissance, significant inventions began to occur almost every decade. During the Industrial Revolution, major technological achievements occurred every few years. Today, significant discoveries are published every year. The scientific and engineering journals are crowded with reports of new technological developments every month.

The reason for exponential growth in technology is clear. The more that is known, the easier it is to discover new knowledge. Knowledge builds on itself. The rate of discovery depends on what has been discovered before. Until mathematics is understood, engineering must proceed by trial and error. Until the spinning and weaving processes were automated, the production of thread and cloth were tedious and labor intensive activities. Until the motions of the sun, moon, and planets were carefully measured, the laws of gravity could not be formulated. Until electricity was understood, electric lights, refrigerators, and television sets were impossible. Until the formalization of computation and the discovery of semiconductors, the modern computer could not exist.

The rate of discovery also depends on the number of scientists and engineers working to extend the frontiers of knowledge. Technological progress does not simply happen. It is the direct result of investment of time and money in research and development, and in the quantity and quality of education and training of scientists and engineers. This means that the rate of technological progress can be accelerated by increasing the level of investment in science and technology. Technological progress is not an "exogenous variable" as many economists would have us believe. It is directly related to economic policy, and specifically to investment policy.

Denial Is Not a River in Egypt

Unfortunately, the disruptive nature of exponential growth in technology and the potential impact on total factor productivity is largely ignored by the economic and political establishment. The conventional wisdom is that technological progress is sporadic, unpredictable, somewhat mysterious, more or less linear, and not much affected by economic policy or the rate of investment. However, this is demonstrably false. The rate of technological progress is directly proportional to the number and quality of the people working in the fields of science and engineering. And the number and quality of researchers is a direct result of the amount of money invested by society in research and development, education, and training of scientists, engineers, and workers. The rate of technological progress is not an exogenous variable. It is directly proportional to the rate of investment and hence subject to control by economic policy.

The reluctance to appreciate the potential of exponential growth is sometimes reinforced by the widespread belief, even among educated people, that technology has gone about as far as it can go. The conventional wisdom is that "everything has limits. Nothing can continue forever." But the exponential growth of technology shows no sign of slowing down, much less stopping. If anything, it is speeding up. Experiments and engineering design cycles that used to take years, now take months. What used to take months, takes days. What used to take days, now takes only minutes. What used to be impossible, is now only difficult. What used to be difficult, is now easy.

The conventional wisdom assumes that sustained economic growth in a mature economy is limited to about 3 percent. There are many that fear that economic growth in the industrialized West will decline due to an aging population and a slowing human birth rate. All of these predictions are based on the premise that human labor will continue to be the principal factor in production and that technology growth is more or less linear. The reasoning is that if the birth rate declines and fewer young people enter the labor force, then economic growth must decline. However, if intelligent machines can achieve the perceptual, cognitive, and decision-making powers of humans, then a growing population of robot workers can more than compensate for a declining population of human workers. In fact, the growing number of robot workers will compete with human workers over a declining number of jobs.

Another form of denial assumes that because early predictions of information technological progress were wrong, or overly optimistic in the past, those predictions will remain invalid forever. For example, early predictions by artificial intelligence researchers grossly underestimated the complexity of intelligent behavior in the real world, and after a flurry of optimism, the field was widely discredited. As a result, most people's fear of losing their job to a robot has subsided from the level of anxiety during the 1980s. But since those early predictions, there have been six orders of magnitude of growth in computational power. A million times increase in anything makes a qualitative difference. In addition, knowledge from neuroscience is growing rapidly regarding how the brain actually works. These advances have brought us to the point where serious efforts at reverse engineering the human brain will probably succeed within a decade or two, and almost certainly before the midpoint in this century.

Unfortunately, this scenario runs outside the boundaries of conventional wisdom, and is therefore not taken seriously. Current economic models are based on past experience. But in a field that is expanding exponentially by two orders of magnitude per decade, what was true in the past will not be true in the future. Economic models are derived from statistical data collected from the past. Until now, it has always been the case that the rate of economic growth depends on the number and quality of human workers. The current model of free market capitalism is based on the assumption that human labor is the principal factor in production. It is not prepared to deal with the possibility that human labor will cease to be the primary source of wealth production in the future.

Under the current model, the middle class relies on the sale of their labor as their primary source of income. The problem is that the economic value of labor is being constrained by the rapidly increasing capabilities and falling cost of advanced automation. What labor can earn is based on supply and demand. If the supply of workers is rising due to growing world population, while the demand for workers is falling because of improved productivity, the market value of labor can only go down. Add to this a rapid increase in the number of low-cost robot workers, and you have the potential for a disastrous collapse of middle-class income.

Recent experience with high unemployment and a jobless recovery should be a clue that the capitalist system does not need as many human workers as it did in the past. In the future, businesses will be able to produce all the goods and services that consumers have money to buy with far fewer workers than the number of people looking for jobs. This suggests that unemployment may remain high for a very long time, perhaps indefinitely.

Current economic theory simply does not consider the possibility that robot labor might replace human labor as the primary source of economic growth. Only the science fiction community has taken this idea seriously. Unfortunately, the dominant theme in the sci-fi world has been a dark vision of massive unemployment such as depicted in Kurt Vonnegut's *Player Piano*,[136] or worse, of artificial life run amuck as suggested by Mary Shelley's *Frankenstein*,[137] or movies of violence such as *Terminator*. Even in the relatively few stories that contain friendly robots such as Isaac Asimov's *I Robot* or the *Star Wars* movies, there is rarely, if ever, serious consideration

---

136   Vonnegut, Kurt. *Player Piano*. New York: Dell Publishing, Random House, 1952.
137   Shelley, Mary. *Frankenstein*. Calgary, Canada: Qualitas Publishing, 1818.

given to how the economic system could create and distribute wealth to the human population in a world where most of the work is done by robots. Almost nowhere have the economic mechanisms needed to avoid the issue of technological unemployment been addressed. The means by which the entire human race might enjoy an aristocratic lifestyle based on robot labor are seldom, if ever, considered.

Only Peoples' Capitalism offers a plan for an ownership society designed to exploit the potential of a robot revolution for the benefit of all humankind.

CHAPTER 6 |||||||||||||||||||||||||||||➡
The Energy Enabler

IN ORDER TO SUSTAIN A rapid rate of economic growth, eliminate poverty, and provide the wealth needed to realize the vision presented in chapter 2, the world is going to have to come up with new sources of clean energy. A continued dependence on fossil fuels is problematic for several reasons. The most immediate of these is the enormous transfer of wealth from the US and other Western Democracies to the Middle East and other countries in the world that are not our friends. This is an immediate threat to our national security, and it saps money from investments in our future.

A second problem is global warming caused by CO_2 pollution. Melting of the glaciers and polar ice is the most visible result, but more importantly, warming of the oceans will increase the frequency and intensity of hurricanes and typhoons, and produce more severe flooding in many regions due to increased evaporation from the warmer oceans.

A longer-term problem is the limited supply of fossil fuel. Within the next half century, the world's supply of oil will be mostly gone, and oil will no longer be an economically viable source of fuel for lighting, heating, air conditioning, and transportation. Supplies of natural gas and coal will last somewhat longer, but eventually sources of energy other than fossil fuels will have to be developed. Among the possibilities are hydroelectric, wind, solar, geothermal, nuclear fission, and hydrogen fusion.[138]

Hydroelectric power requires dams on large rivers, and the number of

---

138   Jacobson, Mark Z. "Review of Solutions to Global Warming, Air Pollution, and Energy Security, *Energy and Environmental Science,* Department of Civil and Environmental Engineering. Stanford, CA: Stanford University, 2008. *http://www. rsc.org/Publishing/Journals/EE/article.asp?doi=b809990c.*

suitable rivers that have not already been exploited is small. Collecting wind energy requires large wind turbine farms that are not suitable for many locations—and the wind does not always blow. Geothermal is not practical except in places where the Earth's crust is thin and volcanic activity is near the surface. Nuclear fission has drawbacks in terms of safety and nuclear waste. Hydrogen fusion poses enormous technical problems that have yet to be solved, and may never be.

Solar

The most promising long-term, non-nuclear, non-fossil energy source is the sun, which of course is a giant hydrogen fusion furnace with enough fuel to last for another four billion years or so. The sun is about 93 million miles away, which is fortunate. Much closer and the oceans would boil. Much farther away and they would freeze. Every day the sun sheds more than enough radiant energy on the earth to supply all of humanity's needs, and will do so for the indefinite future. Unfortunately, this energy is thinly spread over the entire surface of the earth, and most solar energy is simply reflected back into space or reradiated as heat. Some solar energy is absorbed by evaporation of seawater producing clouds, rain, snow, and wind. A small fraction of this power can be captured by electrical generators in dams and wind turbines. A small but significant amount of solar energy is captured by plants through photosynthesis.

At sea level on a clear day when the sun is directly overhead, the energy density of sunlight is about 1.0 kilowatt per square meter. Even if you could capture all of this energy, one square meter of surface area would provide only enough energy to power a small hair dryer. Capturing a commercially valuable amount of solar energy thus requires a large capture area.

Averaged over a twenty-four-hour period, the energy density of sunlight at sea level on the equator is about .4 kw per square meter. This translates to about 10 kilowatt hours (kwh) per day per square meter of area. Current commercial solar cells cost about $600 per square meter and capture only about 20 percent of the energy falling on them. Thus, it takes about 5 square meters of real estate and $3,000 worth of solar cells to generate 10 kwh per day.

Solar cells take a large amount of energy to manufacture. At current efficiencies, it takes one to two years for a conventional solar cell to generate as much energy as it took to make it.[139]

---

139   Green, M. "Thin-Film Solar Cells: Review of Materials, Technologies and Commercial Status," *Journal of Materials Science: Materials in Electronics,* 18 (October 1, 2009):15–19.

However, once installed, solar cells produce electrical power at zero cost. Over a ten-year period, $3,000 worth of solar cells can generate a total of 36,500 kwh. Averaged over a decade, this amounts to only about eight cents per kwh. By comparison, electricity from conventional power plants currently costs about twelve cents per kilowatt-hour in the continental United States and about twice that in Hawaii.[140] Thus, when the installation cost of solar power is amortized over several years, electricity from solar photovoltaics can be cost-competitive with electricity from coal-fired power plants in many parts of the world. [141]

In the future, technology will improve and costs will fall. Solar cell technology is progressing rapidly, and the cost of electricity from solar cells is falling by about 5 percent per year. Analysts predict that solar energy will become commercially cost-competitive with conventional grid-supplied electricity in the 2015 time period in those areas of the world where sunlight is plentiful and conventional electricity is relatively expensive.[142] "Grid-parity for solar power by 2015" is first among the Recovery Act's (a.k.a. Obama's Stimulus Package) Four Investment Goals.[143]

Solar thermal generators are an alternative to solar cells for collecting energy from the sun. These systems use mirrors to concentrate sunlight on a liquid medium that drives a turbine to generate electricity. These are complex systems that require large fields of movable reflectors that track the sun across the sky. A group of four demonstrator solar thermal generators in the California desert are claimed to have a combined capacity of 310mw.[144]

The big problem is to scale up from relatively small experimental installations to compete head-to-head with power from fossil-fuel-fired plants. An enormous financial investment will be required to capture solar energy in sufficient quantities for use in heating cities, generating electricity, and fueling cars, trucks, planes, and ships. Currently, solar power amounts to only about .05 percent of the total energy used in the United States.[145] Scaling up the solar power industry to compete with fossil-fuel-fired plants presents a historic

---

140   Energy Information Administration, Form EIA-861, "Annual Electric Power Industry Report."

141   Wynn, Gerald. *Solar Power Edges Towards Boom Time*. London: Reuters, October 19, 2007.

142   BP Global Power website http://www.bp.com/genericarticle.do?categoryId=9013609&contentId=7005395.

143   Grunwald, M. "How the Stimulus Is Changing America." *Time Magazine*, September 6, 2010.

144   Next Era Energy Resources brochure on Solar Electric Generating Systems, http://www.nexteraenergyresources.com/content/where/portfolio/pdf/segs.pdf.

145   Energy Information Administration data, US Department of Energy data available online at http://www.eia.doe.gov/cneaf/electricity/epa/epaxlfilees1.pdf.

investment opportunity. This is the kind of investment in the future that could pay long-term dividends for Peoples' Capitalism PIP investments.

However, solar power is not without problems. One of these is that electrical energy is not easily stored. Electricity must be used immediately when it is generated. The sun only shines during the day, and not so much on cloudy days. Thus, the electrical power generated by solar cells is not synchronized with the demand for electricity which typically rises when the sun goes down. Of course, electrical energy can be stored in batteries, but major advances will be required in battery technology before battery storage becomes practical for commercial electrical power. On the other hand, heat energy can be stored in liquids relatively easily for several hours. This may enable solar thermal generators to store heat energy during the day and draw upon that heat energy when demand is high in the evening.

Another problem with solar power from either solar cells or thermal generators is that huge tracts of land will be required for producing enough electricity from solar energy to compete with conventional power plants. The solar thermal system mentioned above requires an area of 1,500 acres of California desert to generate 310mw. This is an energy density of about .5Kwh per square meter per day. In the deserts, land may be largely unoccupied and unused. However, not everyone lives near the deserts, so moving the energy from where it is collected to where it is consumed will require a new and much more capable power transmission network. And, if solar power is to largely replace fossil-fueled power plants, it is not clear that the deserts are big enough to supply the world's appetite. Eventually, there will be competition between forests, farm land, cities, and solar power collection facilities. The total electrical power generation capacity for the United States for the year 2009 was about 4×10^{12} Kwh. [146] To generate this much solar power would require approximately 2×10^{10} m^2 of real estate, which is on the order of 0.2 percent of the entire land mass of the United States.[147]

Biofuel

Photosynthesis transforms sunlight into biomass that can be converted into biofuels such as ethanol or biodiesel. Photosynthesis is a process by

---

146   Ibid.

147   1 acre = 4000 m^2. The 1500 acre thermal power station in the California desert occupies 6,000,000 m^2. It generates 310Mw for 10 hours per day = 3100Mwh per day/ 6,000,000 m^2 = .5Kwh per m^2 per day x 365 days per year = 182.5 Kwh per year per m^2. The US consumption of electricity for the year 2009 was about 4×10^{12} Kwh. Therefore, to produce this much electricity by solar thermal power would require roughly 2×10^{10} m^2 of real estate. The total area of the United States including Alaska is about 10^{13} m^2.

which carbon atoms are absorbed from CO_2 molecules in the atmosphere by chlorophyll in the leaves of plants, and combined with hydrogen atoms from water supplied by the roots to produce glucose. In the process, oxygen molecules are released into the atmosphere, and energy is stored in the chemical bonds between the carbon and hydrogen molecules in the glucose. The chemical process in photosynthesis can be expressed as

$$Carbon\ Dioxide + Water + Sunlight\ and$$
$$Chlorophyll = Glucose\ and\ Oxygen$$
$$or\ more\ precisely$$
$$6CO_2 + 6H_2O + Sunlight\ and\ Chlorophyll = C_6H_{12}O_6 + 6O_2$$

When the plants (or their fossilized byproducts such as oil or gas) are burned, the process is reversed, and energy that was stored during photosynthesis is released as heat. Oxygen molecules are absorbed from the atmosphere while CO_2 and water molecules are released back into the atmosphere. However, the CO_2 that is released is that which was absorbed during the process of photosynthesis. Thus, biofuels are carbon neutral and there is no net release of CO_2 into the atmosphere so long as no fossil fuels are used in the planting or harvesting of biomass, or in the processing of biomass into biofuel. There is not even any change in the heat balance, since the heat released during burning is the same as what was absorbed during photosynthesis. In contrast, burning fossil fuels releases CO_2 that was absorbed from the atmosphere millions of years ago. Thus, burning fossil fuels produces a net increase in the current atmospheric CO_2.

Biofuels are nearly as energy dense as gasoline or fuel oil, and much less toxic when spilled. They can be safely and efficiently stored and transported by ship, truck, or pipeline. Biofuels can be burned in power plants for generating electricity, or in homes and offices for heat. They can be also used for fuel in cars, trucks, trains, ships, and planes. An advantage of vehicles propelled by biofuel over electric or hybrid cars and trucks is that biofuel vehicles do not require large batteries that contain toxic materials and heavy metals.

The conversion of sugar cane to ethanol is already a commercially profitable enterprise in Brazil, and most cars in Brazil run on biofuel today. In the United States, the conversion of corn to ethanol has been subsidized by the government, and many brands of gasoline contain ethanol. However, a significant amount of fossil fuel is used in the planting and harvesting of corn and sugar cane, and in the conversion of sugar into ethanol. Thus, the current process technologies are not carbon neutral. Not until biofuels alone are used in the production of biofuel, will the entire process become carbon neutral. Estimates are that conversion of cellulose to biofuel will become commercially viable within a decade. This would enable the production of biofuel from grass, wood, and yard waste which would otherwise decompose and release the same amount of CO_2.

Of course, biomass can be burned directly for heat. This has been done in fireplaces, wood stoves, and campfires around the world since humans discovered how to use fire. It is routinely done in waste treatment incinerators that burn trash and use the heat for industrial processes or to generate electricity. It has recently been tested in a conventional power plant by mixing switch grass with coal as a source of fuel. The Chariton Valley Biomass Project[148] has successfully substituted switch grass for up to 20 percent of the coal in a boiler at the Ottumwa Generating Station in Iowa.

However, biofuels cannot be a long-term replacement for oil, coal, and natural gas until several fundamental problems are solved. The biggest problem lies in the vast amount of biomass required to satisfy the global demand for energy. There is a limited amount of unused terrestrial real estate that is appropriate for biomass production. Most of the world's best farmland is already under cultivation for food crops, and using the Earth's remaining forests and wetlands for biomass production is not an environmentally sound solution. Although the amount of land under cultivation can be increased, and increases in agricultural production can be expected from genetic engineering and improved fertilizers, increased acreage will grow ever more costly as population growth transforms farmland into cities and suburbs. Doubling the amount of land under cultivation is probably not possible.

A second and related problem is that if food and fuel compete for the same land and water resources, demand for fuel will drive up the price of food in the marketplace. As long as food and fuel compete for the same inputs, the price of food will be linked to the price of fuel. This does not bode well for a world population that already has significant problems with hunger, or for countries where a rise in the price of food can lead to riots and political instability. Ethanol production from corn has already had a significant impact on price of corn and products made from corn. The effect of rising prices is spilling over into other cereal crops such as wheat, soybeans, and even sorghum.

A third problem is that there is a limited supply of water in most regions of the world not already under cultivation. Major irrigation projects are enormously expensive, and often are destructive to the environment. Water for irrigation is subject to seasonal variations and drought. Water is already a limited resource in many parts of the world, and water for agriculture is in direct competition with water for human consumption in many places. As populations grow, this problem will become increasingly acute.

---

148   http://www.iowaswitchgrass.com/.

Finally, there is a limited growing season in regions more than fifty degrees from the equator.

Until these problems are overcome, fuel from biomass cannot become a long-term alternative to fossil fuel.

Farming the Oceans

One way that biomass production could be dramatically expanded without affecting the supply of farm land or fresh water would be to farm the equatorial oceans. The area of the earth's surface within twenty degrees of the equator is roughly 67 million square miles, or 171×10^{12} m^2. About 70 percent of this area is water. Thus, the equatorial oceans contain about 1.2×10^{14} m^2 of open ocean—an area twelve times the size of the entire United States including Alaska.[149] The equatorial oceans are warm and days are nearly 12 hours long 365 days per year. Within twenty degrees of the equator, the average solar power at the surface of the earth (averaged over twenty-four hours) is about 400 watts/m^2. Thus, the amount of solar energy falling on the equatorial oceans is equivalent to about 4.8×10^{16} watts, or nearly 50 billion megawatts. If only one tenth of one percent of this energy could be captured, it would produce the equivalent of 50 million megawatts, or roughly 3,000 times the average total world energy rate of consumption for all forms of energy, including coal, oil, natural gas, nuclear, and hydroelectric.[150]

The most promising plant species for growing in the ocean is seaweed. There are many varieties of seaweed that cover a spectrum from single cell algae to giant kelp plants. In between these extremes are many varieties of multi-cell algae. Some of these aggregate into floating mats. Other species anchor themselves to rocks in shallow water. Many varieties of seaweed grow aggressively under the right environmental conditions consisting of nutrient-rich water and sunlight. Seaweed has long been harvested for food, fertilizer, cosmetics, medicines, and biotechnology in many places around the world. The Department of Energy's National Renewable Energy Laboratory (NREL) in Golden Colorado has identified over three hundred species of algae as possible sources of biofuel.[151] In countries as far apart as China and Ireland, coastal waters are often covered by thick, floating mats of multi-cell algae. Some species of algae have the capacity to cover vast regions of ocean with a layer of green.

---

149   CIA World Fact Book gives the area of the United States as 9,629,091 m^2 or roughly 10^{13} m^2.

150   International Energy Agency Key World Energy Statistics 2006. Available online at http://www.iea.org/textbase/nppdf/free/2006/key2006.pdf.

151   National Renewable Energy Laboratory of the US Department of Energy website http://www.nrel.gov/.

Algae is under active consideration by researchers, entrepreneurs, venture capitalists, and energy companies as a source of biofuel. Some species of algae store energy in the form of lipids that can be easily converted into oil for biodiesel fuel. Cultivation of algae has potential for replacing significant amounts of fossil fuel because of its fast growth rate and the high oil content of some varieties. Some species of algae are so rich in oil that it accounts for over 50 percent of their dry mass.

One problem with farming the oceans for biomass is that there is limited coastal real estate that is suitable for farming. Natural algae and kelp beds grow best in coastal shallows where nutrients from silt on the bottom are constantly agitated by wave action. The deep oceans are largely devoid of nutrients necessary to grow biomass.

Many forms of seaweed such as kelp must be attached to underwater structures such as rocks on the bottom. This further limits the available acreage. Artificial kelp mooring systems have been developed, but these are vulnerable to storm damage.

One solution to these problems is to build large, floating algae farms consisting of transparent tubes filled with a broth of algae and nutrient-rich water that would be capable of surviving the environment of the open oceans.

Design for an Open-Ocean Algae Farm

The first requirement for a floating algae farm is that it must be able to maintain its structural integrity on the open ocean environment. It must be rigid in the horizontal plane so as to not wrinkle or fold, and flexible in the vertical direction so as to ride smoothly over the top of the waves. It must be capable of navigating out of shipping lanes and away from approaching storms. It must be capable of sinking below the surface in heavy weather and rising back to the surface when the storms have passed.

The design shown in figure 6.1 is one possibility. I make no claim that this is the best or only conceivable design, only that it is illustrative of what is possible and might ultimately become practical. What is important is that this approach exploits the equatorial oceans where sunlight and water are in abundant supply. This is an enormous area that is almost entirely uninhibited. The environmental impact of farming operations in these waters would be negligible. Yet, capturing even a tiny fraction of the solar energy that falls on this region of the earth could enable a very bright future for human civilization.

The design shown in figure 6.1 mimics the structure of a biological leaf built around the hull of a large ship at the base of the stem. There are many tanker ships in the current inventory that could be converted at a minimal cost. Alternatively, ships could be specifically designed for the purpose.

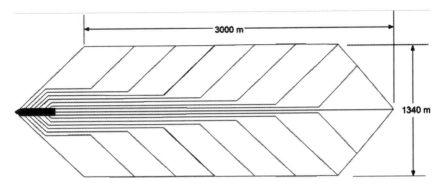

Figure 6.1. Top view of a leaf-shaped algae farm structured around a converted oil tanker. The ship is shown at the left. The outlines indicate fields of transparent tubes filled with a broth of algae and seawater saturated with CO_2 and nutrients.

The area of this farm is 4 million m² or roughly one thousand acres. It probably could be made bigger. The circulation of algae broth flows away from the ship through ribbons of transparent tubes to the fields where it circulates in the sun, and then back to the ship where the algae is extracted to be processed, and nutrients and CO_2 are added to the water before it is recirculated. The transparent tubes contain a series of unidirectional flow valves that cause wave energy to pump the water-algae broth through the tubes and back to the processing ship. As the waves rise and fall, the broth in the transparent tubes will slosh back and forth. Every twenty meters, unidirectional flow valves similar to those in the mammalian heart restrict flow in the backward direction and enable flow in the forward direction.

The fields of transparent tubes circulating the algae broth are supported by a square matrix of pressurized tubes filled with seawater as shown in figure 6.2.

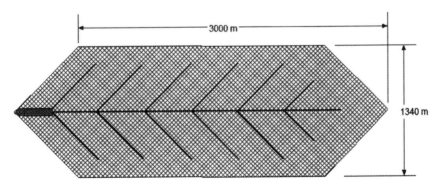

Figure 6.2. A neutrally buoyant truss structure consisting of fabric tubes inflated with pressurized seawater attached to the ship with steel or Kevlar cables. The small squares in this drawing are 20m x 20m.

The pressurized support matrix would be attached to the sides and rear of the ship just below the layer of transparent tubes. Towing forces would be distributed throughout the matrix by a series of steel cables that are illustrated as heavy lines in figure 6.2. A design for a single 20 m x 20 m element of the support matrix is shown in figure 6.3.

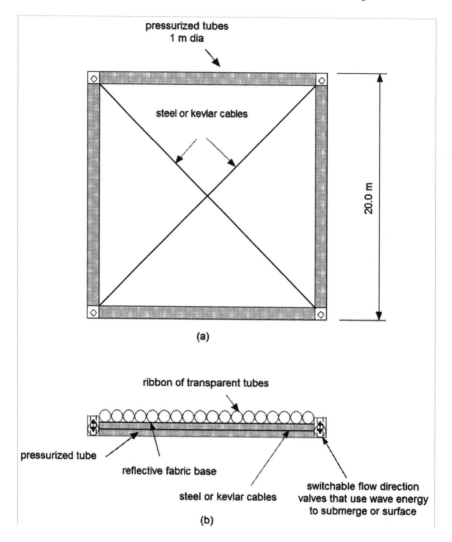

Figure 6.3. A 20 m element of the square structural truss that provides horizontal stiffness and supports the transparent tubes filled with algae broth. (a) is a top view with the transparent tubes and fabric base removed to show the pressurized tubes, corner connectors, and crossed steel or Kevlar cables that provide diagonal structural stiffness. (b) is an end view showing the

reflective fabric base supporting a ribbon of eighteen transparent tubes that circulate the algae broth.

The fabric tubes are one meter in diameter and filled with seawater pressurized to about 3 psi. A pressurized fabric tube of this design will withstand a compressive force of more than twelve hundred pounds before crushing, and would quickly return to a straight configuration if bent or crushed by large wave action. The length-to-diameter ratio of 20:1 enables the pressurized fabric tubes to resist Euler buckling.

The pressurized tube matrix would be neutrally buoyant and would support the transparent tubes just at the ocean surface. The internal pressure in the support matrix would cause it to be stiff in the horizontal plane to prevent folding or bunching up due to the forces of waves, wind, or differential currents, but flexible in the vertical dimension so as to conform to long ocean waves.

The ribbons of transparent tubes would contain a rich broth of nutrients with an atmosphere of CO_2 and O_2 released as a byproduct of photosynthesis. Between the transparent tubes and the support matrix would be a waterproof, reflective fabric base. This would increase the efficiency of algae growth because it reflects sunlight back through the transparent tubes. It also would serve as a barrier to the vertical flow of seawater.

Switchable directional flow valves located in the corner connectors of the 20 m x 20 m elements can be electronically switched open or closed to cause wave energy to pump water upward for submerging, or downward for surfacing. If seawater is pumped up, the structure will submerge. To submerge completely the transparent tubes must also be purged of any gas (CO_2 or O_2) that may have accumulated inside. If seawater is pumped down, the structure will rise to the surface.

Around the periphery of the pressurized matrix would be a floatation collar to prevent breaking waves from pounding the transparent tubes.

It may be necessary from time to time to clean the transparent tubes, as algae may become attached to the inside of the tubes. For this task, a robot can be designed to move through the insides of the tubes and scrub them using energy provided by wave action to power itself.

Concept of Operations

During normal operations the floating algae farm would navigate the equatorial oceans to avoid storms and keep clear of shipping lanes. The primary means of propulsion would be flaps on the underside of the support structure that capture wave energy. Waves cause water beneath the leaf structure to surge back and forth. When the surge is in the forward direction, the flaps fill with water and produce force to push the leaf structure forward. When the surge is in the reverse direction, the flaps collapse and produce no force. These flaps could be differentially deployed so as to steer the leaf in the desired direction. The ship could also use its engines for emergency propulsion to allow the algae farm to navigate in calm seas or to move quickly when necessary to avoid storms or shipping lanes.

The floating algae farm would be visited periodically by tanker ships that would off-load equipment, personnel, and supplies, including fertilizer and liquid CO_2, and up-load processed algae products. The fertilizer and CO_2 would be held in tanks aboard the farm ship, and dissolved in the algae broth to promote rapid growth.

Estimated Production Rate

The structure shown in figure 6.1 covers about 4 million square meters (roughly 1,000 acres). Assuming a production rate of between 2,000 and 20,000 gal/acre/year, a farm this size could produce the equivalent of between 2 million and 20 million gallons of oil for biodiesel per year. This is equivalent to between 36,000 and 360,000 barrels of oil per year, or between 100 and 1,000 barrels of oil per day. At $100 per barrel, this would return between $3.6 million and $36 million worth of product per year per farm.

The current US consumption of oil is about 20 million barrels per day. So it would take about 20,000 algae farms of this size to supply the entire US need for oil. About 80,000 farms this size would be required to supply the entire world's need for oil.[152]

Eighty thousand farms would take up roughly 160 thousand km². The equatorial oceans contain about 120 million km² of empty space. Thus, enough farms to supply the entire world's demand for oil would occupy about .13 percent of the space in the equatorial oceans. If equally spaced, there would be only one algae farm every 27 km (roughly every 16 miles) in each direction. Thus, there would be plenty of room for maneuver, and if necessary, expansion.

Navigation

Hurricanes and typhoons are the biggest threat to open-ocean algae farms. Fortunately, hurricanes are largely confined to summer and fall in each hemisphere, and rarely occur within ten degrees of the equator. Thus, the algae farms can migrate to the northern equatorial oceans between December and May, and to the southern equatorial oceans between June and November.

---

152   Data from the Energy Information Administration, US Department of Energy, http://tonto.eia.doe.gov/country/index.cfm.

An average speed of 1 knot is probably sufficient to stay clear of trouble. This would enable a kelp farm to navigate about 25 miles per day, or 9,100 miles per year.

In the case of an occasional storm that cannot be evaded, the algae farm is designed to submerge to a depth such that storm energy would not damage the structure. This could be done by dumping air from the floatation collar and the transparent tubes, and setting the switchable flow direction valves located in the corner connectors of the 20 m x 20 m elements so that pressure from the vertical motion of waves beneath the leaf structure forces water up through the valves. Thus, with every passing wave, water is pumped from beneath the leaf structure to above it. This causes the leaf to sink to a depth where the wave energy is not stressful to the structure. At that depth, the switchable flow check valves would be closed, and the structure would stop sinking.

As the strength of the waves caused by the storm subsides, the switchable flow check valves can be set so that pressure from wave motion forces water down through the valves. In that configuration, every passing wave pumps water down, from above to beneath the structure. This causes the leaf to rise to the surface where air can be pumped back into the floatation collar. The leaf structure can therefore stabilize itself at any desired depth by using wave energy to pump water either up or down.

It is anticipated that situations that require submerging the algae farm would be rare. Good weather predictions should be able to predict storms far enough in advance for the farm to avoid rough seas without interruption of normal operations. When necessary, the ship could use its engines to maneuver out of the way of storms.

Other hazards may include whales, dolphins, and large fish. These can be kept away with high intensity sonar or electrical fields. Robot patrol boats may be used to defend the perimeter if necessary, and repairs can be made by human divers with helicopter support from a ship.

Environmental Impact

From an environmental standpoint, there are few options better than farming the equatorial oceans for fuel. Hydroelectric dams flood canyons. Windmills can be eyesores. Nuclear plants generate radioactive waste. Biomass grown on land destroys forests and competes with food crops. Solar power facilities occupy large amounts of terrestrial real estate. Open-ocean farms would operate far from land in parts of the ocean that are almost completely unoccupied even by fish and sea mammals. The deep ocean is for the most part a desert, devoid of life.

It may also be practical to grow fish in the transparent tubes. The algae being cultivated for fuel could be mixed with phytoplankton and other species of algae that provide food for fish. The directional flow valves would not harm fish, and the fish could be diverted from the filters that harvest algae. In short, the open-ocean farm might provide the opportunity for growing both fuel and food in an ecologically friendly and sustainable manner for the indefinite future. The technology to make this happen is within reach. Nothing in the above description would require any scientific breakthrough. All that is required is the application of well-known engineering principles and an adequate supply of investment capital.

Open-ocean algae farms may thus provide a significant investment opportunity for the Peoples' Capitalism Personal Investment Program. The long-term return on investment for individual investors could be significant, and the long-term benefits to society in terms of energy independence and reduced dependence on fossil fuels could be even more important.

CHAPTER 7 ‖‖‖‖‖‖‖‖‖‖‖‖‖‖‖‖‖‖‖‖‖‖‖‖‖➡

The Common Defense

I N HIS 1987 BOOK *THE Rise and Fall of the Great Powers*, Paul Kennedy traces the history of national and international power since the Renaissance.[153] His unifying theme is the interaction between economic strength and military strength. Kennedy shows how in every case, the military strength of a nation is a direct reflection of its economic strength. The ability of a nation to prevail in conflicts with others is determined more than anything else by the size and growth rate of its economy. Superior technology is an important but temporary advantage. Superior numbers are important, but not as important as the ability to equip troops in the field with the best weapons and training. Superior geography is important, but not as important as the manufacturing capacity to equip armies with the most effective equipment.

Kennedy's book predicts the relative and absolute decline of the United States due to military overstretch, deficit spending, and slow economic growth. In 1987, these were theoretical problems. Today they are painfully real. One of the lessons that can be drawn from history of the last five centuries is that the great powers use their military power to extend their influence to far-flung regions of the globe. Eventually, these ventures become a drain on the home economy that leads to slow economic growth and relative decline in wealth and power. Today, the United States is immersed in two land wars deep in Asia, in danger of two more wars with Iran and North Korea, and is facing a rising super power in China. Democracy in Pakistan is in danger of being overtaken by Islamic extremists which would deliver nuclear weapons into the hands of terrorists. We also have tens of thousands of troops in Europe,

---

153   Kennedy, P. *The Rise and Fall of the Great Powers*, New York: Random House,1987.

Japan, and South Korea. We are spending as much on our military as all the other nations in the world combined. At home, we are deeply in debt, have slow economic growth, and face the prospect of an extended period of high unemployment.

There are, of course, ways to reverse this negative trend. The United States still has an enormous advantage in many different areas. But we are not using our advantage to good purpose. To balance our budget, we could raise taxes enough to pay for the government benefits that are politically popular. But that would require we all contribute our share, not just the rich. The deficit is running in the neighborhood of $1.5 trillion per year, which amounts to about $5,000 per year for each United States citizen. To balance the budget we would need to raise taxes by $5,000 per person just to pay for what the government is spending. The national debt is $14 trillion. To retire that would require roughly $46,000 per person. On the other hand, we could cut spending for pensions, Social Security, and medical care, but that would adversely affect many people, and plunge many into poverty, homelessness, and hunger. And the pain will be felt most intensely by the poor and elderly. We could cut defense spending, but that would weaken our ability to maintain our global military commitments.

Or we could grow our way out of this situation. If we were to increase our investment rate by 10 percent of GDP, it would generate millions of new jobs. The economy would begin to grow at a 6 percent rate. If we increased our investment rate by 20 percent, the GDP would grow at a 9 percent rate. In that case, GDP would double every 8 years, quadruple in 16 years, and grow by nearly 74 times in 50 years. Tax revenue from this rate of economic growth would enable us to balance the federal budget, fully fund Social Security, Medicare, and Medicaid, lower tax rates, and maintain a strong military, all simultaneously.

If we were to institute Peoples' Capitalism, the Personal Investment Program would provide the necessary investment to produce 6 percent to 9 percent economic growth. The Personal Saving Program would provide the necessary savings to maintain price stability. Technological innovations in advanced materials, nanotechnology, clean energy, automation, and robotics would boost productivity enough to support a 6 percent to 9 percent rate of economic growth.

In the near term, employing underused capacity in the current economy could support this rate of economic growth. In the mid-term, increased productivity would support this rate of growth. And in the long term, the technology of human-level machine intelligence would effectively expand the labor force with robot workers. This could support economic growth rates in the neighborhood of 9 percent per year indefinitely. Once human-level machine intelligence is achieved, the production of goods and services could grow with the number of intelligent machines employed. The economy might grow by .7 percent for every 1 percent added to the labor force by robot workers. And consumer demand would grow with the increased productive output, because the profits earned by the automated businesses would pay dividends to the general public. Rapid economic growth would enable industry to produce more wealth, and widespread ownership would assure a prosperous citizenry. These rates of economic growth would assure a strong industrial base capable of supporting a strong military and political presence in the world.

The National Security Imperative

The advent of human-level machine intelligence will almost certainly be achieved first for military applications. As has been the case for so many technological advances throughout history—starting with the forging of metal weapons, the domestication of horses and elephants, the development of the compass and the clock, sailing and steam-powered ships, nautical charts and maps, the airplane, the computer, Earth satellites, the transistor, the integrated circuit, radar, lasers, nuclear energy, and the Internet—the technology was either developed specifically for the military, or heavily financed by government for military applications. Certainly, the earliest and most significant source of funding for intelligent systems technology has been the military.

The job of the soldier, sailor, and airman is inherently dangerous. At the point of contact with the enemy, life expectancy is short. Particularly in operations in urban terrain, where soldiers must move down narrow streets under sniper fire from rooftops and windows, where enemy fighters mingle with the civilian population, where troops must enter buildings that are booby-trapped with explosives and infested with suicidal defenders, military operations become bloody and personal. Improvised explosive devices and roadside bombs are a serious threat to conventional armies fighting native insurgents on foreign territory.

The ability to project force from a distance lies at the heart of military weaponry. It began with the sling, spear, and bow and arrow, and continued with the rifle, the machine gun, the mortar, the howitzer, the battleship, the airplane, the aircraft carrier, and the guided missile. The Predator drone Unmanned Aerial Vehicle (UAV) is the latest embodiment of this principle.

The Predators are flown under remote control by human pilots, and the weapons are controlled by human gunners that are far from the battlefield. Pilots and weapons technicians in command centers in Colorado can see, track, and destroy moving targets in Afghanistan, Pakistan, and Somalia.

These are only the first in a series of increasingly sophisticated intelligent robotic systems that will enter the military weapons arsenal over the coming decades. The Global Hawk is a pilotless surveillance plane that can loiter at 60,000 feet altitude for thirty-two hours, and fly from the United States to Australia without refueling. Both Boeing and Grumman have developed pilotless fighter-bombers that can take off from and land on aircraft carriers, provide close air support to ground troops, engage in air-to-air combat, and destroy enemy targets on the ground such as trains, trucks, tanks, anti-aircraft radar, and missile batteries.

The first and foremost rationale for artificial intelligent systems in military operations is to reduce casualties and increase military effectiveness. As robot intelligence grows, the human warrior will become more of a commander, and teams of robots will undertake the more dangerous frontline combat roles. Technology is under development that will enable robots to perform tactically significant tasks in complex urban environments, and interact with humans in close-knit teams.

A second rationale for military interest in intelligent robots is readiness and training. Robotic weapons can be stored in containers on ships and bases around the world, and can be activated on a moment's notice to fight fiercely and skillfully with suicidal intensity. Military robots can be programmed with skills and abilities that are translated from textbooks and tested in the field in combat training exercises. Training can be performed in both physical and computer-generated environments against both human and computer-generated opponents. Once useful military tactics, techniques, and procedures are encoded in software, they can be duplicated at minimal cost and stored in computer memory ready to be activated upon command. Depending on operational priorities and rules of engagement set by human commanders, military robots can be programmed to be more or less aggressive, more or less stealthy, and more or less willing to accept risk.

The primary technical challenge is perception. Significant research efforts are being focused on autonomous perception technology. Advanced image processing and sensor fusion algorithms are beginning to enable robot vehicles to see and understand visual images from sensor suites on robot vehicles. Data from color and infrared cameras can be combined with input from laser range imaging cameras and structured light projectors to build high-resolution

3-D models of the environment. Objects can be segmented, their attributes measured, their behavior characterized, and their relationships with other objects determined. Next-generation robots will be able to focus attention on objects and events of interest, and reason about situations and episodes. They will possess the ability to detect patterns, understand relationships, and recognize trends that can be assigned value and meaning. They will be able to evaluate what is happening in the world. This will give them the ability to make decisions amid uncertainty, predict the consequences of various potential courses of action, adapt to changing circumstances in a dynamic and unpredictable world, and generate behaviors that conform to rules of engagement and commanders' intent. This technology will increase the size and effectiveness of our military forces, and reduce the cost of training and readiness.

Military robots can be large vehicles that carry heavy weapons or transport supplies. They can be mid-sized vehicles that carry light weapons, or perform as general purpose "mules" to carry supplies for dismounted troops. They can be small robots that search tunnels, caves, and buildings for enemy fighters and booby traps. They can be tiny robots that fly just ahead and above ground troops, or perch on buildings, atop utility poles, or in trees to observe activity in busy streets or markets. They can be disguised as rocks, or tree bark, or parts of buildings.

Unmanned vehicles can operate on land, in the air, on the water, or undersea. In the air, UAVs such as the Predator and Global Hawk are already highly developed and widely used. Unmanned bombers and fighter aircraft are being developed and tested that can fly in formation with manned aircraft in a variety of combat missions, and perform the most dangerous tasks, such as attacking anti-aircraft missile sites and providing close air support to ground operations.

On land, small robots are used to work with bomb squads to dismantle mines and improvised explosive devices (IEDs.) Larger and more intelligent ground robots are used to patrol ammunition dumps and military warehouses, or to carry special-purpose sensors to detect and mark mines and IEDs. Robots are under development that can stand guard, patrol fence lines, deliver supplies, drive trucks in convoys, and serve as lead vehicles in patrolling urban streets and alleys.

On the water, intelligent robots are being developed to patrol harbors and provide security around manned ships. Undersea, intelligent robots can patrol and spy, or lie on the bottom for months or years waiting for instructions. They can carry explosive charges and behave as smart mines or torpedoes.

Unmanned undersea robots can operate at much greater depths than manned submarines. They can operate in teams where one robot emits acoustic pings and others passively listen for echoes. They can be deployed from manned submarines and later recovered to be refueled and report their findings. Future developments will enable undersea robots to mimic large fish, extract energy from ocean waves, and monitor traffic in and out of ports for months or years at a time.

Intelligent robots can be controlled by programs that encode the tactics, techniques, and procedures that are taught to human soldiers, sailors, and airmen. In the near future, they will have the ability to learn from instructors in training exercises along with human soldiers. Eventually, these robots will exhibit a capacity for situation analysis, common-sense reasoning, decision-making, and real-time planning that approximates the capabilities of human warriors.

Robots have numerous inherent advantages over humans on the battlefield. The most obvious is that when they are killed, no grieving families have to be informed, and no dead bodies have to be flown back home. When robots are wounded, there is no need for humans to risk their lives to drag them to safety, and no need for medics or field hospitals to treat their injuries—only repair technicians. When they are lost behind enemy lines, there is no need for dangerous rescue operations. Robots can sustain higher shock, blast, vibration, G-forces, and noise levels than humans. They can survive greater extremes in temperature, and are not vulnerable to biological or chemical warfare agents. Although they require fuel, they don't need food, water, toilets, or sleep. And finally, they don't require expensive training, housing, medical care, rehabilitation, or retirement benefits. As the developmental costs are amortized over a large number of units, intelligent robots will be far less expensive to train and maintain in a state of readiness than human warriors.

The ability to endow robotic ground vehicles, aircraft, submarines, and surface ships with human-level artificial intelligence will change the way wars are fought. In his book *Wired for War,* P. W. Singer suggests that intelligent battlefield robots may turn out to be more significant than the atomic bomb in the history of warfare.[154] Singer argues that robotics will alter not merely the tactics of war, but the very identity of those who fight it.

---

154   Singer, P. W. *Wired for War: The Robotics Revolution and Conflict in the 21ˢᵗ Century,* New York: Penguin Press, 2009.

The Long-Term Consequences

Unfortunately, battlefield robotics will only temporarily impact the military balance between the great powers. While technology is difficult to develop, it is also difficult to keep proprietary. The classic example of this principle can be seen in the history of nuclear weapons. It took a national effort of enormous magnitude to develop the first atomic bomb. Yet, only four years after the Americans dropped atomic bombs on Hiroshima and Nagasaki, the Soviet Union exploded its first atomic bomb.[155] The United States exploded the first hydrogen bomb in 1952, and the Soviet Union duplicated that feat only one year later.[156] This suggests that once the technology to build human-level machine intelligence exists anywhere, it will rapidly be duplicated by every major power with the will and ability to commit comparable scientific, engineering, and financial resources to the task.

Thus, while the race toward human-level intelligence for robots is primarily driven by the military's desire to possess the most powerful and effective weapons, the most important long-term impact will be on productivity and economic growth in the nation's industrial capacity. For as is clear from Kennedy's *Rise and Fall of the Great Powers*, economic strength is the most important factor in military power. In the long run, military strength is ultimately determined by economic strength, and the two cannot be separated. The relative military strength between any two powers is determined by the relative strength of their economic systems.

This raises a national security concern relative to the Chinese. Today, the United States' economic strength as measured by GDP is roughly four times greater than that of China. However, the Chinese economic growth rate is roughly 10 percent per year, while US economic policy is purposely designed to limit growth rate to 3 percent. Figure 7.1 shows the predicted performance of the Chinese and US economies under two scenarios over the next thirty years. The first scenario is where the US continues to grow at 3 percent and the Chinese economy continues to grow at 10 percent. In this scenario, the United States' economic strength will remain greater than that of the Chinese for only nineteen more years. After that, the Chinese economy will race ahead of the US economy at an accelerating rate. This does not bode well for America's future in this century.

---

155   Rhodes, Richard. *The Making of the Atomic Bomb*, Simon and Schuster, New York: Simon and Schuster, 1986.

156   Rhodes, Richard. *Dark Sun: The Making of the Hydrogen Bomb*, New York: Simon and Schuster,1995.

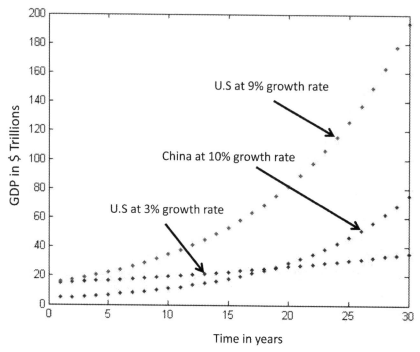

Figure 7.1. The US GDP under 3 percent and 9 percent growth compared to China's GDP growing at 10 percent over a thirty-year period. In 2008, the Chinese GDP was $4.3 trillion and the US GDP was $14.6 trillion. The bottom two curves show the present trajectory (10 percent growth in China vs. 3 percent in the US). These curves predict the Chinese GDP will surpass the US GDP in nineteen years. However, if the US were to boost its growth rate to 9 percent, the US would maintain its lead for more than a century.

The second scenario in figure 7.1 shows the US growing at 9 percent during the same time frame. In this case, the US maintains its lead over China for more than a hundred years.

Figure 7.2 shows the same two scenarios over a fifty-year period. The power of compound interest is even more dramatic in this figure.

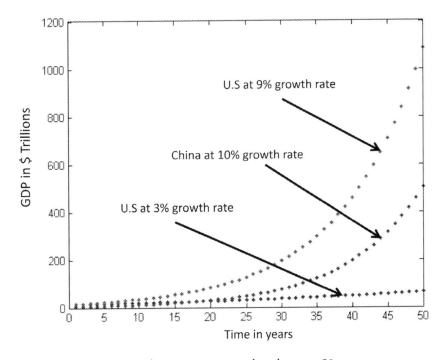

Figure 7.2. The two scenarios played out to 50 years.

If the US and China maintain their current respective growth rates for 50 years, the US GDP will be $62 trillion while the Chinese GDP will be $500 trillion, or 7.4 times larger than the US. On the other hand, if the US were to boost its growth rate to 9 percent, the US GDP would be $996 trillion in 50 years, or 2 times greater than the Chinese. It should be noted that even a 1 percent difference in economic growth matters in the long run. Even if the US economy grows at 9 percent, if the Chinese economy continues to grow at 10 percent, they will pass us in about 130 years. The ratio of US-to-Chinese GDPs would slowly fall due to the difference between 9 percent and 10 percent growth rates, but the absolute difference would continue for about 130 years.

It is seldom mentioned that the United States' rate of investment is consistently lower than the world average.[157] The United States' rate of investment has averaged less than 20 percent of GDP since shortly after

---

157   World bank data available online at http://www.google.com/publicdata/explore?ds=
       d5bncppjof8f9_&ctype=l&strail=false&nselm=h&met_y=ne_gdi_totl_zs&hl=en&
       dl=en#ctype=l&strail=false&nselm=h&met_y=ne_gdi_totl_zs&scale_y=lin&ind_y
       =false&rdim=country&idim=country:USA&tdim=true&hl=en&dl=en.

the end of World War II. This is more than 2 percent less than the world average, and far less than that of the Chinese. Since 1993, the Chinese rate of investment has averaged 40.6 percent of GDP, and in 2007 was over 44 percent.[158] This explains why the Chinese rate of economic growth has average about 10 percent per year for the past two decades, while the US growth rate has struggled to average 3 percent.[159] Only if we match the investment rate of our major competitors are we likely to prosper as a nation economically, and remain the premier military superpower for the rest of this century.

It thus is a matter of national security that the United States increase its investment rate substantially—at least by 10 percent of GDP and preferably by 20 percent. Unfortunately, given the current set of monetary and fiscal tools available to the government, it is not conceivable that any combination of interest rates, tax cuts, and Keynesian stimulus programs could stimulate anything near a sustained increase in investment of 10 percent of GDP, much less 20 percent of GDP, in the United States. Furthermore, it is difficult if not impossible to imagine how to form a political consensus around a policy that could increase the investment rate by 10 percent of GDP or more so long as ownership of the means of production remains almost the exclusive property of the big banks, insurance companies, rich individuals, and large corporations. Only if the ownership of capital assets is widely distributed, so that everyone benefits from increased investment, are we likely to find a political consensus to end the long-entrenched US policy of underinvestment.

Energy Independence

One obvious place where increased investment could make the United States more prosperous and secure would be in developing technology to eliminate the need for buying foreign oil. If we used some of our increased investment to deploy technology such as solar power or farming the oceans for biofuel, it would produce a major shift in the geopolitical landscape and dramatically improve our national security. The current dependence of the Western nations on oil from the Middle East, Russia, and Venezuela tends to prop up undemocratic regimes and undermine our credibility on human rights. It

---

158   World bank data available at http://www.google.com/publicdata/explore?ds=d5bnc
       ppjof8f9_&ctype=l&strail=false&nselm=h&met_y=ne_gdi_totl_zs&hl=en&dl=en
       #ctype=l&strail=false&nselm=h&met_y=ne_gdi_totl_zs&scale_y=lin&ind_y=false
       &rdim=country&idim=country:USA:CHN&tdim=true&hl=en&dl=en.

159   World bank data available at http://www.google.com/publicdata/explore?ds=d5bn
       cppjof8f9_&ctype=l&strail=false&nselm=h&met_y=ny_gdp_mktp_kd_zg&hl=e
       n&dl=en#ctype=l&strail=false&nselm=h&met_y=ny_gdp_mktp_kd_zg&scale_
       y=lin&ind_y=false&rdim=country&idim=country:CHN:USA&tdim=true&hl=en
       &dl=en.

has produced an enormous transfer of wealth from the United States, Japan, and Europe to countries that are not our friends. Nations in the Middle East have many grievances against the West that stretch back to the Crusades. In modern times, countries in the Mid-East have been invaded by the British, the Germans, the Russians, and the Americans. Many Muslim sects have deeply held religious beliefs that make them intolerant of infidels, of which we are.[160] Muslim countries in the Mid East also have a proud and lengthy history that extends back to the empires of Babylonia under Hammurabi and Nebuchadnezzar, the empire of Persia under Cyrus the Great and Darius the Great, and the Ottoman empire under Suleiman the Magnificent, a.k.a. His Imperial Majesty Grand Sultan, Commander of the Faithful and Successor of the Prophet of the Lord of the Universe. It is not a good strategy to be sending these people billions of dollars a day in cash for oil. It is even greater folly to send our soldiers to fight religious extremists on their home territory which they consider to be holy ground. This inevitably results in death of civilians and destruction of homes which only intensifies their hatred of us regardless of our good intentions.

The primary reason given by Osama bin Laden for his jihad against the United States was the presence of US troops on the ground in Saudi Arabia. One of the biggest recruiting tools of al Qaeda is the presence of US troops in Iraq and Afghanistan, and Predator drone attacks in Pakistan. If we didn't need the oil, we could leave that part of the world to the people who live there. We could provide humanitarian aid when requested, and execute counter-terrorism operations when necessary, but we would not need large expeditionary forces deployed on their territory, and we would not need to prop up friendly but corrupt governments that are hated by the local people. No one wants foreign troops in their country. We would not be happy if Iraqi troops were kicking down doors in the United States and killing American citizens. Why should we be surprised when we are not greeted as liberators when our helicopter gunships fire missiles into their villages?

If we want peace, we need to make ourselves independent of Mid East oil, stop sending them our wealth, and get our soldiers out of their cities and off of their land. The only long-term solution to terrorism is an attractive economic alternative. We cannot kill them all, nor should we want to. We need to show nations in the Mid East and around the world how it is possible for all people to participate in the economic and social benefits of modern

---

160   Historically, almost all religions think themselves to be the true religion, and
      consider all those that do not believe this to be heretics at best, and infidels at
      worst. This tendency is most common and intense among the poor and uneducated.
      The ultimate solution is to provide all people with a decent income from ownership
      of capital so they have some hope for a good life in this world.

technology. We need to demonstrate how even the poor nations of the world can alleviate poverty and suffering by broadening the ownership of the means of production. We need to provide an alternative to the rapidly growing slums and refugee camps of the world that provide breeding grounds for revolutionaries and terrorists.

Ultimately, our national defense will be most reliably secured by introducing to the world a new form of capitalism by which poverty can be eliminated, and all the world's peoples can share in the ownership of the means of production. Better weapons and bigger armies are only a temporary solution. In the long run, it is economic strength that determines national power.

CHAPTER 8 ‖‖‖‖‖‖‖‖‖‖‖‖‖‖‖‖‖‖‖‖‖‖‖‖➡

Benefits to Humankind

"We the people of the United States, in order to form a more perfect union, establish justice, insure domestic tranquility, provide for the common defense, promote the general welfare, and secure the blessings of liberty to ourselves and our posterity, do ordain and establish this Constitution for the United States of America."

Preamble to the United States Constitution

"The progressive development of peoples is an object of deep interest and concern to the Church. This is particularly true in the case of those peoples who are trying to escape the ravages of hunger, poverty, endemic disease and ignorance; of those who are seeking a larger share in the benefits of civilization and a more active improvement of their human qualities; of those who are consciously striving for fuller growth."

POPULORUM PROGRESSIO
Encyclical of Pope Paul VI on the Development of Peoples

RAPID ECONOMIC GROWTH COUPLED WITH widespread ownership of the means of production would satisfy both the goals of the US Constitution and the concerns of the Catholic Church. Rapid economic growth stimulated by increased investment would create many new jobs. It would reduce unemployment and provide opportunity for many to escape from poverty by getting a job, building productive capacity for the future.

Rising income from ownership of capital assets would gradually eliminate the ravages of hunger, poverty, disease, and ignorance, and bring the benefits of civilization to all of humankind. Widespread ownership of productive assets would strengthen democracy. Individuals would not need to depend on government benefits, and would not be vulnerable to radical movements. Widespread ownership would support justice and insure domestic tranquility. Individuals would be economically and physically secure in their persons and their property. This would create an era of peace, prosperity, and economic justice that would promote the general welfare and secure the blessings of liberty to ourselves and our posterity.

Peoples' Capitalism provides mechanisms for achieving rapid economic growth and widespread ownership of capital in a manner such that everyone benefits and no one loses. The Personal Investment Program (PIP) would generate a massive infusion of new money into the economy targeted exclusively for capital investment. In the near term, this would spur economic growth and create enough jobs to eliminate unemployment. It would create new customers and markets for goods and services, and provide many new business opportunities. It would cause the stock market to soar and would provide investment funding for new businesses. In the long term, the PIP would provide income from ownership of capital assets to individuals. Everyone —poor, middle class, and rich—would receive an exponentially growing supplemental income, independent of jobs, generated by an exponentially growing portfolio of capital assets. Eventually, as income from investments rises, workers will begin to drop out of the labor force voluntarily. This will leave jobs available for those who want to continue working.

The Personal Savings Program (PSP) would maintain price stability so that rapid economic growth could be sustained indefinitely. Mandatory PSP savings would give individuals the added security of substantial personal savings accounts. PSP savings would also give the Federal Reserve a new and much more effective tool than monetary policy for controlling inflation and maintaining the value of the currency. This would free up the use of interest rates and monetary policy for encouraging savings and attracting foreign investment.

Under Peoples' Capitalism, those who are ambitious, hardworking, and successful would continue to be rewarded for their achievements. Rapid growth would provide many opportunities for individuals to get rich—even extremely wealthy—by dedicating their lives to business, finance, science, engineering, medicine, or law. Increased investment would stimulate the economy to grow at 6 percent to 9 percent per year. Jobs would be plentiful and deficits would shrink. Financing for small businesses would be easy to get. Opportunities for entrepreneurship would abound. And consumers would have plenty of money to spend, so that market demand would be strong and steady.

Under Peoples' Capitalism, every human being would be recognized as a potential customer. Everyone has needs and wants. But wants and needs are not enough. It takes income to transform wants and needs into consumer demand in the market. For most people, income is tied to jobs or retirement benefits. No jobs => no income => no consumer demand. This means that if all businesses minimize labor costs, there will be a shortage of customers. And if there is a shortage of customers, businesses will need fewer workers —which leads to fewer jobs—which leads to fewer customers—which leads to fewer jobs …. This is a deadly spiral that the invisible hand of lassie-faire capitalism cannot fix.

Peoples' Capitalism provides a way to break out of this spiral. In the short term, increased investment would create demand for workers to build new machines, factories, plants, and transportation facilities. This would create demand for parts, tools, materials, and workers. Investment spending creates jobs that generate income that increases market demand, and causes businesses to hire and invest even more. In the long term, investment spending pays for itself and creates income for investors from dividends, interest, or rent. Investments made through the Peoples' Capitalism PIP pay dividends to average people who use it to buy the new products and services that their investments produce. Dividend income for everyone leads to more customers, which creates more profits, which creates more dividends, which creates more customers …. This is a virtuous spiral.

Investment also generates productivity growth which enables businesses to produce more and better products and services at lower cost with fewer workers. Under the current version of capitalism, the economic benefits of productivity are limited to the owners and the currently employed workforce. But productivity reduces the need for workers, which unless the economy is growing rapidly causes unemployment to grow, and the number of people living in poverty to increase. However, if access to credit for investing were broadened to cover the poor and middle class, the income benefits of productivity will be translated into dividends for everyone. In that case, a growing income stream from capital investments would supplement, and eventually supplant, wages and salaries. This would enable many workers to voluntarily drop out of the labor force to retire early, to go back to school, to spend more time with family, or to start a business. Middle-class families will once more be able to get along on one worker's salary. Would-be entrepreneurs will be empowered to pursue their dreams. Would-be inventors will have more leisure to pursue innovative ideas. Workers will get more vacation days. Life will get less hectic and pressured. The economy will grow rapidly even while the number of jobs declines and the number of people choosing to work in the traditional labor force declines as well.

The PSP would make everyone into a capitalist. Everyone would have income from a growing ownership share of the means of production. Increased investment and saving would spur economic growth, and the distribution of wealth increasingly through ownership of capital would provide a degree of financial security to every citizen. This would provide freedom from hunger, freedom from fear, and freedom to pursue the blessings of liberty to the entire citizenry.

Economic Growth and Domestic Tranquility

Today the political debate in the Unites States is highly partisan and divisive, primarily because of slow economic growth and persistent unemployment. "Where are the jobs?" has become a political battle cry. The middle class is under great pressure. Middle and lower incomes are flat or declining. Jobs are scarce. Poverty is increasing. Unions have lost their power to force owners to give workers a fair share of the profits. Polls show that over 60 percent of the population feels the United States is in decline. But the owners of capital are doing very well. The upper levels of the income hierarchy have seen their incomes increase, and the nearer to the top, the greater the percentage increase. The gulf between the rich and the poor is the largest since 1929.[161] The conventional wisdom is highly pessimistic. Krugman writes of an era of Diminished Expectations.[162] Almost everyone agrees that the US economy is not doing well, and the political fight is over what should be done about it.

The Republican Party adheres to the doctrine of small government and low taxes. They want to cut government entitlements and reduce benefits for the middle class and poor, while cutting taxes and providing incentives for the rich to invest. The claim is that this will create jobs. The trouble is that private investors don't invest unless they think their investments will pay off. If there is excess capacity and sluggish demand, investors don't invest, and businesses don't hire. So the unemployment remains high and the economy grows slowly. The Democratic Party wants to expand government benefits, and use government spending to stimulate the economy. The trouble is that very little of government spending goes for capital assets that pay dividends in

---

161   Sherman, Arloc and Chad Stone. *Income Gaps Between Very Rich and Everyone Else More Than Tripled In Last Three Decades, New Data Show*. Washington, DC: Center on Budget and Policy Priorities, June 25, 2010 http://www.cbpp.org/cms/index.cfm?fa=view&id=3220.

162   Krugman, Paul. *The Age of Diminished Expectations*. Cambridge, MA: MIT Press, 1994.

the form of tax revenues. Raising taxes enough to pay for continued stimulus spending is politically untenable. The resulting impasse leads inevitably to deficit spending and a growing national debt.

Peoples' Capitalism Solutions

Among the biggest challenges for the mature democracies in America, Europe, and Japan is how to reconcile globalization, declining economic growth, aging populations, and growing mountains of government debt. The Peoples' Capitalism solutions to these problems are the following:

1. The solution to globalization is to give the average citizen an opportunity to acquire an ownership share in the means of production. When you are an owner, it doesn't matter where the production facilities are located, so long as your dividends grow steadily and arrive on time.

2. The solution to declining economic growth is to increase the investment rate and boost productivity. Investment spending stimulates the economy in the short term, pays dividends in the mid-term, and generates productivity gains that support long-term growth.

3. The solution to an aging population is to: a) improve productivity so that fewer workers can produce more wealth, and b) distribute the increased wealth to an aging population through ownership of capital assets that pay dividends. Dividends from investing would provide economic security for the elderly and handicapped, and enable them to afford good health insurance.

4. The solution to the growing mountains of government debt is to increase the rate of economic growth so that tax revenues rise with the economy.

Unfortunately, no one today is talking about the possibility that we could grow our way out of the current economic predicament. The idea of growing the economy fast enough to solve our future economic problems was a popular notion during the Reagan administration.[163] But the "supply-side" economic theory of the 1980s has been discredited. Only the most hard-core devotees still believe (against overwhelming evidence) that cutting taxes pays for itself in tax revenue. But there are many who still believe that tax cuts are the secret to economic growth, despite the fact that the Reagan tax cuts never generated the promised amount of economic growth, and after a decade in place, the

---

163   Gilder, George. *Wealth and Poverty*. New York: Basic Books, 1981.

Bush round of tax cuts turned a record surplus into a huge deficit and presaged the worst economic collapse since 1929.

Supply-side economics was based on the premise that reducing regulations and cutting taxes primarily on the rich would induce a higher rate of saving and investment, and thereby grow the productive capacity. In practice, however, much of the tax relief for the rich was spent on bigger houses, fancier cars, designer clothes, consumer electronics, luxury items, yachts, and vacation homes. Only a small fraction of supply-side tax cuts actually ended up being invested in improved productive capacity. During the Reagan and Bush-41 years, the investment rate actually fell from 22.2 percent of GDP in 1979 to 16.2 percent in 1991.[164] Most taxpayers invest only a tiny fraction of their income. Many live from paycheck to paycheck, and many more are deeply in debt for consumer items. Most individuals invest nothing in research and development of new technology, contribute very little to maintaining the nation's infrastructure, and invest far less than what is necessary to generate rapid economic growth. Hence the predicted economic growth never materialized.

An Era of Austerity

Michael Gerson in a May 19, 2010, column in the *Washington Post* editorial page entitled "The Era of Austerity" described the pain and social upheaval in Greece, Spain, Portugal, and Romania that he attributes to "decades of welfare-state comfort and years of Keynesian stimulus spending." His analysis is that "resentful debtor nations are being forced by their creditors into tax increases and spending cuts that are painful, unpopular—and just beginning." Austerity measures that reduce government jobs, cut salaries, and reduce pensions cause angry citizens to riot in the streets as governments become insolvent and economies collapse. Gerson suggests that Greece and Romania are only the most recent examples. He predicts that Great Britain and eventually the United States will come to the same fate. He is pessimistic because of the difficulties of cutting government spending and fears that increasing taxes will lower economic growth and weaken job creation. These kinds of predictions are all based on the assumption that nothing can be done to increase economic growth. In an article published in the *Harvard Business Review* in August 1997 entitled "How Fast Can the Economy Grow?" Paul Krugman wrote that:

---

164   World Bank data available online at http://www.google.com/publicdata/explore?ds=
       d5bncppjof8f9_&ctype=l&strail=false&nselm=h&met_y=ne_gdi_totl_zs&hl=en&
       dl=en#ctype=l&strail=false&nselm=h&met_y=ne_gdi_totl_zs&scale_y=lin&ind_y
       =false&rdim=country&idim=country:USA:CHN:JPN&tdim=true&hl=en&dl=en.

"Most economists believe that the US economy is currently very close to, if not actually above, its maximum sustainable level of employment and capacity utilization. If they are right, from this point onwards growth will have to come from increases either in productivity (that is, in the volume of output per worker) or in the size of the potential work force; and official statistics show both productivity and the workforce growing sluggishly. So standard economic analysis suggests that we cannot look forward to growth at a rate of much more than 2 percent over the next few years. And if we - or more precisely the Federal Reserve - try to force faster growth by keeping interest rates low, the main result will merely be a return to the bad old days of serious inflation."[165]

Robert Samuelson in a 2010 syndicated column wrote about "The Age of Austerity" that has already arrived in Europe and is destined for the United States wherein governments are cutting social spending and raising taxes. His diagnosis is that "the welfare state and the bond market have collided, and the welfare state is in retreat." His take-away line was, "the Age of entitlement was about giveaways, the Age of Austerity will be about take-backs."[166]

Steven Pearlstein in an article entitled "Wage Cuts Hurt, But We Need Them" argues that the only way out of the current recession is for workers to accept "creative new wage structures" (a.k.a cuts in wages) to bring labor costs in the United States into line with wages overseas. He says "this will better spread the burden of getting the economy back into balance" by reducing the amount that we consume.[167] In other words, American workers are going to have to compete with Chinese, Korean, Taiwanese, and Indian workers in the global labor market, so get used to declining wages.

The failure of supply-side economics to produce rapid economic growth had produced widespread disillusionment, even despair, especially on the right. However, the real lesson to be learned from the failure of supply-side growth theory is not that faster economic growth is impossible, but that cutting taxes is not the way to achieve it. Most tax cuts never make it into investment. Most go for increased consumption. The oft-stated premise that the people can make better use of their money than the government is simply not true. Government spending on roads and bridges, research and development, utilities, firemen, police, public health, and safety are actually

---

165   Krugman, Paul. "How Fast Can the Economy Grow?" *Harvard Business Review*, July/August 1997.

166   Samuelson, Robert J. "The Age of Austerity." *Washington Post*, October 11, 2010, A17.

167   Pearlstein, S. "Wage Cuts Hurt, But We Need Them." *Washington Post*, October 13, 2010, A13.

investments in infrastructure that pay dividends to society as a whole. Taking that money and giving it to taxpayers in the form of lower taxes mostly just increases consumer spending. Tax cuts mainly stimulate consumer demand. This temporarily increases economic activity, but the jobs created are based on consumer products and services that don't earn profits or return dividends to the consumers. Even tax cuts to businesses are often not spent on investment unless the business leaders expect consumer demand to rise. Otherwise, they simply use tax cuts to increase profits to the business owners. Thus, most tax cuts produce very little long-term economic growth. If tax cuts are financed with government borrowing, once the spending is done, all that is left is the debt.

The power of investment to make the economy grow rapidly is undeniable. The problem with both Keynesian stimulus and supply-side economics was that in practice they do not significantly increase investment in capital assets. The secret of making capitalism really work is to create money and put it directly into investments that pay back more in dividends than the cost of the investments.

Almost everyone agrees that investment is the key to economic growth. The current political argument is over how to make it happen. Peoples' Capitalism makes it simple. The Federal Reserve issues credit (i.e., creates money) that is directly invested by individual citizens in capital assets that pay dividends and trade on the open market. The amount of credit should be sufficient to produce a real economic growth rate of 6 percent to 9 percent per year, and the investment managers must be seasoned professionals that are competitively selected and judged on their performance in return-on-investment.

Under Peoples' Capitalism, increased investment would produce the advanced technology for improving productivity and sustaining rapid economic growth. It would promote the general welfare by creating a society in which everyone is financially secure. It would secure the blessings of liberty by enabling everyone to acquire sufficient capital assets to provide a livable income.

The Future of Capitalism

There is serious cause for concern regarding the future of capitalism in democratic countries. In a democracy, politicians cannot get elected unless they reflect the attitudes and beliefs of their constituents. Constituents typically want lower taxes and more benefits for themselves and their families. This creates a strong bias toward budget deficits and borrowing to finance deficits. As a result, most democratic governments have promised their constituents future benefits that exceed estimates of future revenue. There are three possible ways this can be solved. One is to cut benefits. A second is to raise taxes. The third is to grow the economy fast enough so that the budget can be balanced without either cutting benefits or raising taxes.

Unfortunately, the notion that we can grow our way out of this predicament is now considered naïve. This is ironic because we live in a time when technological progress is faster than ever before, and technology is the primary source of productivity growth which is the ultimate source of economic growth. Given the state of modern technology and the exponential growth in productivity-enhancing developments in automation and robotics, we should be experiencing rapid and accelerating economic growth. The problem is that the West in general, and the United States in particular, are not investing enough to exploit these technological advances.

In contrast, countries in Asia, notably China, South Korea, Singapore, and India, are experiencing rapid economic growth. China's economy has grown around 10 percent almost every year since 2003. Even during the economic crash of 2008 it grew at 9 percent. India has grown faster than 8 percent since 2003, and grew at 6 percent during 2008. These growth rates are easily explained by the rate of savings and investment in these two countries. China's investment rate has been above 40 percent of GDP since 2003, and India's investment rate has been above 30 percent since 2004. These investment and growth numbers are unheard of in the United States, except for the period just before and during WWII when the US government poured money into investments in industrial capacity for the war effort. Since 1950, the US investment rate has dropped below 20 percent of GDP and the economic growth rate has averaged 3.9 percent. Since 2000 with tax rates at an all-time low, the US growth rate averaged only 2.5 percent.[168]

The way to increase economic growth is to increase the investment rate, and do it directly, not through tax cuts or government spending. The way to insure that consumers will have the purchasing power to buy the new products and services produced by the increased investment is to channel investments through individual citizens so that they reap the profits from their investments and use their profits to buy the new products that are produced by their investments.

To grow rapidly, an economy needs both a high rate of investment and consumers with rising incomes to buy the products. The Asian economies are growing rapidly because they have a high rate of savings and investment at home, and are selling their products abroad. They rely on the export market to consume the increased production output because they don't have a viable

---

168   World Bank data available online at http://www.google.com/publicdata/.

way to distribute wealth to their domestic consumers other than jobs, and they don't need a lot of additional workers to generate all the output they can sell. The Western economies are growing slowly because their savings and investment rates are low and their domestic markets are saturated. In both East and West, there are plenty of poor people that are potential consumers, but most of these are unemployed and can't find work. People without jobs don't have enough income to buy the products that could be produced. As a result, we have people in need and unused productive capacity side by side. In the poor, we have people who need all kinds of goods and services. In unemployment, we have unused workers who could be employed for producing wealth to meet the needs of the poor. But, because we have no means by which income can be created for the poor, they remain both poor and unemployed. And the economy stagnates.

The way to grow our way out of the looming debt crisis is to invest more in capital assets that pay dividends, and distribute those dividends to the general population so that they can buy the products that the new capital assets would produce. An increase in investment of 10 percent to 20 percent of GDP would generate an economic growth rate of 6 percent to 9 percent. This would make the solvency problems of Social Security, Medicare, and Medicaid disappear. More importantly, it would lead to a more tranquil and prosperous society. And it would strengthen our national security and international prestige and influence. It would demonstrate to the world that the enormous benefits of capitalism can be obtained without the inherent unfairness of an oppressive class structure of owners at the top, with workers in the middle, and the unemployed poor at the bottom. Peoples' Capitalism would build an income floor at the bottom, but impose no ceiling on the top. With the introduction of Peoples' Capitalism, the workers and poor could become the owners without taking anything from the current owners. Within a few decades, everyone would be prosperous and economically secure, while opportunities to become rich would abound. Such an economic system would enable Jeffersonian democracy for the post-industrial era.

If people around the world were financially secure, it is hard to imagine how oppressive or tyrannical governments could arise, or if already entrenched, survive. If the world were to become a place where everyone had income from an ownership share in the means of production, slums would disappear. Families would no longer need to live on city trash dumps with children picking through garbage. They would live in comfortable rental apartments and eventually in their own homes as their stock portfolios grow and dividend incomes increase. In communities where people are secure in ownership of the means of production, revolutionary ideologies and religious extremists cannot flourish.

This is not to say that financial security would solve all problems. As long as there are humans, there will be conflict and controversy. There will always be wrongs to be righted, but the wrongs would not be of the kind that cry out for violent retribution. There will always be tribal conflicts. One only needs to look at the fans at football or soccer[169] games to see how deeply tribal loyalties affect human behavior. But violent extremist movements require masses of desperately poor people living on the edge of starvation, in tents, shacks, and slums, without sanitation or clean water, without any semblance of security or justice, without political representation, often caught between fighting armies, and with no hope of escape from destitution. If oppressed masses of the poor no longer existed, reformers would adopt less destructive methods. In communities where people are prosperous, disputes tend to be solved in court, not at the point of a gun, or the tip of a knife, or in the blast of a bomb.

Environmental Preservation

Many have expressed concerns regarding the environmental impact of rapid economic growth. Rapid economic growth in China and India are causing significant problems of air and water pollution from factories and power plants. Air pollution from coal-fired power plants and automobiles has recently become a serious problem in China as millions of Chinese have been able to afford air conditioning, electric appliances, and cars instead of bicycles as their main mode of transportation. But China is rapidly moving toward more efficient coal-fired power plants, wind turbine technology, and rapid rail-transportation systems. Their rapid economic growth has enabled them to aggressively pursue green technologies along with rapid economic development.

The worst pollution occurs in the poorest parts of the world, not the richest. It is the poor that live in squalor and dump raw sewage in the streets, streams, and rivers. The worst environmental pollution occurs in slums and refugee camps where people cannot afford good sanitation, clean water, or decent housing. More prosperous communities have clean water, adequate sewage treatment facilities, regular garbage collection, good schools, and modern health care facilities. The rich can afford the cost of pollution control and environmental preservation. The poor cannot.

Modern technology has the capacity to create new sources of raw materials, more efficient power plants, and cleaner forms of energy. A clean and healthy environment is expensive. It is a luxury that is not available to the poor. Clean air and water, decent housing, adequate medical care, and modern education

---

169   Called football everywhere but in the United States.

cost money. They can be provided to everyone only if we grow the economy fast enough and make the income stream broad enough so that everyone can afford to pay for these services.

If we create a new economic system that provides a source of income from advanced industrial production to all, we can have the wealth to preserve and protect the environment. Farming the oceans for biofuel could eliminate most, if not all, the air pollution and CO_2 emissions from burning fossil fuels. But this would take a very large investment in a new technology. At least initially, biofuel will be more expensive than coal, oil, and natural gas. It will not be practical in the near term unless we generate wealth at a faster rate. Similarly, setting aside land for parks and wildlife habitat is expensive. It requires significant amounts of money to buy and protect wilderness preserves.

Population

Many of the problems of pollution derive from the sheer number of people in the world. The human population cannot continue to grow indefinitely. The earth is a finite body, and it is the only habitable planet in the solar system, and hence in our universe. Unless we find a way to slow and eventually stabilize population growth, the number of human beings will overwhelm the environment. So far, the only proven, non-coercive approach to reducing population growth is to increase prosperity so that people can be secure in old age without depending on a large number of children. The prosperous countries of Europe have actually stabilized their population, as have the Japanese. The Chinese have done so as well, but with draconian measures that violate human rights.

The problem with slowing population growth is that aging populations require significant economic growth to finance decent health care for people in their declining years. Without rapid productivity growth, slowing population growth is problematic. Eradicating disease, educating the poor, providing clean water, building modern sewage treatment facilities, and providing economic security for the elderly as the world population ages will require huge amounts of money that will not be available unless we increase the economic growth rate. Only rapid economic growth will generate the wealth needed to provide a decent standard of living for all of the world's people in a world with a stable population.

Life without Work

In 1995, Jeremy Rifkin published a book entitled *The End of Work*. In it he documents the decline of the global labor force as a factor of importance in the production of wealth. He traces this to the "extraordinary gains in productivity which has allowed companies to produce far more goods with far fewer people."[170] He documents how labor is losing its ability to demand a fair share of the new wealth that is created. He further shows that the number of jobs being generated is far less than the number of jobs being eliminated by advanced automation. And this is happening worldwide! Jobs being eliminated are in the agricultural, manufacturing, construction, transportation, and service sectors. The workers being displaced are farm workers, factory workers, construction workers, truck drivers, and service workers. Rifkin argues convincingly that education cannot be the solution for the large majority of these people. The jobs being created are in the knowledge sector. This is essentially a boutique sector consisting of relatively few entrepreneurs, executives, investors, and software developers.

Rifkin is quite pessimistic. In his chapter on "A More Dangerous World," he analyzes the consequence of growing unemployment. He shows a rather alarming correlation between unemployment and violent crime.

Unfortunately, while Rifkin describes the problem in graphic detail, he offers no plausible solutions. He suggests a volunteer economic sector of the economy that could occupy the time and talents of displaced workers, but he ignores the main point of working—income. People do not go to work to fill empty hours. They go to work because that is how they earn money. Volunteerism provides occupation for people that already have a comfortable income. It is not a substitute for income. Peoples' Capitalism offers the mechanisms for generating income independent of work. Peoples' Capitalism would make the volunteer option feasible.

Some have suggested that without jobs to occupy their time, people would become couch potatoes, drunks, or drug addicts. But this has not been the case for generations of rich people who have more than enough money to live without working. The rich often work at jobs that interest them and which allow them to feel productive. Many who are rich work in business, law, medicine, science, or politics. Many do volunteer work. Many have elaborate hobbies. When people have enough income to live without working, they typically find interesting and challenging ways to spend their time. They travel, pursue education, do home improvement, participate in sports, and attend the theater. It is the unemployed poor that sit at home and watch TV, join street gangs, or take drugs. This is because they don't have the money or education to do anything more interesting or productive with their time. If the poor had income, they would find interesting and productive ways to occupy themselves.

---

170   Rifkin, Jeremy. *The End of Work*. New York: Penguin Books, 1995: xviii.

World Peace and Stability

The only long-term hope for world peace and stability is rapid economic growth that provides income for the billions of poor and unemployed so that they can become consumers of the goods and services that modern technology and capitalism is capable of producing.

Of course, in the short term, domestic peace and order must continue to be enforced by local police, and foreign threats must be countered by military force. The world is a dangerous place, and short-term tactics are necessary to enable long-term solutions to take effect. So a strong security force is necessary, both domestically and internationally. For at least the next few decades we will continue to need large police and military forces with technological superiority. But, even that requires a strong economy so the demands of security do not inhibit rapid economic growth.

In the long term, however, the best hope for both domestic tranquility and international peace is to provide economic justice so that everyone has a degree of financial security. Well-to-do neighborhoods do not require a massive police presence. Prosperous nations settle their differences through negotiations. Certainly, most of the world's ills could be cured if every individual in every country had a source of income that allowed him or her to be both physically and financially secure. In societies with economic justice, grievances over past wrongs, and disputes over land and property can be solved in court, without gunfire or bombs. In a world with rapid growth and economic justice, ethnic and religious differences would no longer give rise to violent confrontations. People would be able to focus on a growing prosperity for themselves and their families.

Peoples' Capitalism could be used to introduce economic justice in almost any country in the world. Almost every country has a central bank and a set of regional banks that could grant loans to individuals for investment in mutual fund stocks. These mutual funds would invest in the most profitable enterprises, both in the local country and in the world, and their stock would trade on the open market. Thus, the people of any country could have access to credit for investing in the most profitable companies in the world, and the most attractive government bonds in the world. Any central bank could issue credit to its citizens for investing in high-quality capital assets. In extremely poor and/or corrupt countries, the credit might be issued by the World Bank or the International Monetary Fund. Since the PIP provides loans to individuals in a manner that would be relatively simple to monitor, the potential for fraud and corruption could be minimized. There would, of course, be a need for significant and meaningful oversight to assure responsible behavior on the part of investment managers, and to verify the identity of the individuals participating in the program. The central bank and individual investors would hold joint title to the mutual fund stocks until the debt is retired. Loan repayments to the central bank would be deducted from the profits before residuals are paid to individuals. These transactions could be made transparent and easy to audit.

There are, of course, risks in investing. Sometimes investments fail to return a profit. However, if the portfolio managers are well trained, competitively selected, and properly motivated, and if the investment portfolios are adequately diversified, they can be expected to generate profits with great reliability. Good investments in high-technology industries typically produce profits in excess of 25 percent. Returns on investment to mutual fund shareholders in a rapidly growing economy routinely achieve 8 percent.

Prospects for Prosperity

If the United States were to take the lead in adopting Peoples' Capitalism, the peoples of the world could see the US economy growing at 6 percent to 9 percent per year, with every citizen acquiring a growing portfolio of capital assets. This example would be so attractive that people would demand that their own governments introduce some version of Peoples' Capitalism. America would clearly be seen as the land of opportunity. The US would truly become the "shining city on a hill" that inspired President Reagan.

Poverty in America would disappear within a generation. Jobs would be plentiful and well paid. As dividends from investment portfolios rise, everyone would be on track to a livable income from ownership of capital assets. Onerous jobs would have to pay well, because no one would be forced by economic necessity to do them. This would drive up the labor costs for less desirable occupations, and provide a strong economic incentive to automate them. Thus, robots would quickly fill the dull, dangerous, and unpleasant jobs. More attractive jobs would pay less, but would be filled by volunteers and people that enjoy working. In the long term, most workers would be people that don't really need the money, but work for the extra income and the satisfaction that comes from contributing to society and being successful in the work place.

The solution to poverty and slow economic growth in any country is to increase investment in productive assets, and provide income to the poor so that they become consumers. This would enable capitalism to produce the wealth needed to provide clean water, adequate food, decent housing, good sanitation, affordable health care, and free education to everyone.

In Conclusion

Free market capitalism has failed to produce goods and services for the poor because the poor have no income. The poor flock to the cities looking for work because labor-intensive farming is no longer economically viable. But there are no jobs in the city to absorb the flood of people looking for work. Modern industry doesn't need hundreds of millions of uneducated and unskilled workers. Businesses can produce all the products that they can sell with the number of employees that they have now. In fact, most businesses could probably get rid of 10 percent of the people they have on the payroll, and it would improve their bottom line. Modern industrial technology is reducing the number of workers needed in the manufacturing sector around the world, and job reductions in the service sector are not far behind. There are barely enough new jobs being created in the cities to employ all the people that already live there. Millions of immigrants only become unemployed slum dwellers.

So long as income is tied to employment, the unemployed will be left out of the capitalist economy. The unemployed don't have the income needed to generate consumer demand, so industry does not produce products to meet their needs. As fewer and fewer workers are required, the number of consumers with disposable income will decline. As consumer demand slows, businesses cut back on production and lay off workers. Economic growth slows while human needs grow. We are entering an era where productivity is reducing the need for labor in all sectors of the economy—not just in farming and manufacturing, but in construction, transportation, and service sectors as well. The new jobs being created are highly specialized and not suited for masses of unemployed. Fewer workers leads to lower consumer demand, which reduces the need for labor still further. There is no guarantee that the number of new jobs created will match the number of old jobs lost. In the past, this just happened to be true. In the future, it will not. Unemployment is already a major problem worldwide. This is only the beginning of a growing crisis. The solution is to create an income stream that is independent of jobs for everyone. Only this will enable the vision set forth in chapter 2 of this book.

Education is often cited as the solution to the technological unemployment problem, but it is not. There are many good reasons why people should be educated. Education is good and should be encouraged by all means. But education is not the answer for the great masses of humanity without jobs. To begin with, education is a long-term and very expensive investment. The people migrating from the countryside to the megacities are desperately poor.

They are literally living on the edge of survival. They need income today, not tomorrow.

But in the long term, educating the masses will not solve the unemployment problem. It will only lead to educated unemployed masses. Education enables individuals to get ahead of their less-educated contemporaries. Education is a passport to better-paying jobs. But the fact is that the modern industrial system doesn't need masses of additional workers, even highly educated workers, to produce all the goods and services that can be sold in the market. Every year, at an exponential rate, technology is making human labor less and less essential for production. This is not a positive trend so long as income for most people depends on having a job. Peoples' Capitalism would solve this dilemma. In the near term, increased investment would grow the economy and create jobs. In the long term, it would provide income to everyone independent of jobs.

Peoples' Capitalism would also provide many secondary benefits. Income from ownership of capital assets would reverse the migration of the poor from the countryside to the slums of the cities. It would enable people to remain in rural communities and still receive a livable income. Income to family farms, small towns, and villages would enable rural economies to provide local jobs for carpenters, masons, teachers, farmers, and health care providers. People would not have to migrate to the cities in search of income or opportunity. Rural economies would thrive as local residents receive rising income from ownership of capital assets. People in the countryside could afford to build water and sewer systems, schools, and hospitals. People in small villages could afford better housing, indoor plumbing, electricity, refrigeration, heating, and air conditioning. If individuals in the countryside had income, there would be no reason to flock to cities in search of jobs. Life in villages and small towns would be much more attractive than living in crowded city slums.

Individual ownership of capital assets would provide a degree of dignity and freedom to all individuals. In many societies, this would dramatically improve the lives of women. Wives would no longer be entirely dependent on their husbands for money. Junior family members would be financially able to go to school, start a business, or pursue a career in science or the arts. People would be financially secure in their old age. Pressures for large families due to fear of destitution in old age would subside. As individual wealth and financial security grows, population growth tends to slow. In the long run, income from ownership of capital assets would lead to a sustainable economic system where per capita wealth grows, but the total population is stable or even declines. This would produce a world where concern for environmental preservation is supported by the economic resources needed to reduce pollution and set aside land for parks and habitat for endangered species. Nations would be financially able to develop renewable, carbon-neutral energy sources, and to build environmentally friendly cities and towns.

There is little doubt that the science and technology to achieve a much higher rate of economic growth exists—or at least could exist given the right investment decisions by government and industry in research and development. Over the next few decades there will be rapid advances in materials and manufacturing processes. There will be advances in nanotechnology and microelectronic circuitry, advances in medicine and health care, advances in computational hardware and software, and advances in communications and networking. Perhaps the most significant advance will be the development of intelligent machines with perceptual and cognitive capabilities that rival those of the human brain and mind.

Technology is approaching the Kurzweil Singularity.[171] Human-level machine intelligence is only a decade or three away, depending on funding levels. Advances in neuroscience are revealing how the brain works. Advances in computer science and robotics are revealing how to build intelligent systems.[172] Computational power is approaching that of the human brain. There are good scientific and engineering reasons to believe it will soon be possible to reverse-engineer the human brain.[173]

The implications are enormous. Human-level perception, reasoning, decision-making, planning and control on laptop-class computers will change everything. It will enable an explosion of productivity that will improve quality of products and services, reduce costs, and increase profits in every sector: manufacturing, transportation, construction, mining, health care, education, national defense, and environmental restoration. Commensurate advances will occur in materials, drugs, chemicals, and medicine. Cures will be found for disease. Aids to the disabled and elderly will be developed.

It is clear, from the current pace of technology development, that by the year 2050 most goods and many services will be provided by automated machines, intelligent transportation facilities, and Internet-based retail suppliers. Under the current capitalist system, this will eliminate millions of jobs and will put significant downward pressure on wages and salaries. So long as income depends on jobs, as jobs are lost, consumer demand falls. As demand falls, businesses cut jobs, and unemployment becomes a persistent problem. If we stay on the current path under the current capitalist system,

---

171   Kurzweil, Ray. *The Singularity Is Near: When Humans Transcend Biology.* New York: Penguin Group, 2005.

172   Albus, James S. and Alexander M. Meystel. *Engineering of Mind: An Introduction to the Science of Intelligent Systems.* New York: John Wiley & Sons, 2001.

173   Albus, James S. "A Model of Computation and Representation in the Brain." *Information Sciences,* Vol. 180, (2010): 1519–1554.

the rate of economic growth will continue to be restricted by monetary policy to 3 percent. The owners of capital will continue to do well, but the gap between the rich and the remainder of society will grow, and poverty will steadily increase.

The bottom line is that technological innovation and the fundamental principles of capitalism are providing us a two-edged sword. One edge is the opportunity to create a new era of peace and prosperity. The other edge is the threat of massive unemployment, growing poverty, and civil unrest. Technology is creating a generation of intelligent machines that will eventually be capable of producing and delivering all of the goods and most of the services that modern society needs and wants with very little human labor. Capitalism provides the mechanism of individual ownership of corporate assets that could provide an income stream for all humankind. This could make democracy and peaceful coexistence a viable way of life for every country in the world. But some things will have to change. We will have to find a better way to finance investment and savings, and we will have to provide access to credit for investment to everyone, so that we all can become capitalists with a reliable and secure source of income from ownership of capital assets.

Peoples' Capitalism provides a way to do this. Within a single lifetime, we could create an Everyperson's Aristocracy in America for ourselves and our posterity. By the end of the century, an Everyperson's Aristocracy would be a possibility anywhere in the world. Will it happen? It depends on whether we perceive the possibility and act to make it happen.

BIBLIOGRAPHY

Albus, James S., et al. *4D/RCS Version 2.0: A Reference Model Architecture for Unmanned Vehicle Systems, NISTIR 6910.* Gaithersburg, MD: National Institute of Standards and Technology, 2002.

——— "A Model of Computation and Representation in the Brain." *Information Sciences* 180 (2010): 1519–1554.

——— "A theory of cerebellar function." *Mathematical Biosciences* 10 (1971): 25–61.

——— *Peoples' Capitalism: The Economics of the Robot Revolution.* Kensington, MD: New World Books, 1976.

——— "Reverse Engineering the Brain." *International Journal of Machine Consciousness* 2, no. 2. (2010): 1–19.

——— "Toward a Computational Theory of Mind." *Journal of Mind Theory* 0, no. 1 (2008): 1–38.

Albus, James S. and Anthony J. Barbera. "RCS: A cognitive architecture for intelligent multi-agent systems." *Annual Reviews in Control* 29, issue 1 (2005): 87–99.

Albus, James S. and Alexander M. Meystel. *Engineering of Mind: An Introduction to the Science of Intelligent Systems.* New York: John Wiley & Sons, 2001.

Anthony, David. *The Horse, the Wheel, and Language: How Bronze-Age Riders from the Eurasian Steppes Shaped the Modern World.* Princeton, N.J.: Princeton University Press, 2007.

Ashenfelter, Orley C. and Richard Layard, eds. *Handbook of Labor Economics*. Amsterdam: Elsevier Science B. V., 1986.

Ashford, R. and R. Shakespeare. *Binary Economics: The New Paradigm*. Lanham, MD: University Press of America, 1999.

Ashford, R. and D. Kantarelis. "Capital Democratization." *The Journal of Socio-Economics*. Vol. 37, No. 4, 2008: 1624–1635.

Ashford, R. "Binary Economics – An Overview." Syracuse, NY: Syracuse University Law School, 2006. Accessed May 30, 2011. http://papers .ssrn.com/sol3/papers.cfm?abstract_id=928752.

——— "A New Market Paradigm for Sustainable Growth: Financing Broader Capital Ownership with Louis Kelso's Binary Economics." *Praxis: The Fletcher Journal of Development Studies*, XIV (1998): 25–59.

——— "Louis Kelso's Binary Economy." *Journal of Socio-Economics* 25 (1996): 1–53.

——— "The Binary Economics of Louis Kelso: The Promise of Universal Capitalism." *Rutgers Law Journal* 22, no. 3, (1990).

Bailey, Brian J. *The Luddite Rebellion*. New York: New York University Press, 1998.

Baruzzi, A., C. Franzini, E. Lugaresi, P. L. Parmeggiani, *From Luigi Galvani to Contemporary Neurobiology: Contributions to the Celebration of the IX Centenary of the University of Bologna. Bologna, 27–28 September, 1988 (FIDIA Research Series)*. New York: Springer, 1990.

Benz, Karl. 1886. Vehicle with gas engine operation. The Imperial Patent Office 37435, filed January 29, 1886, and issued November 2, 1886.

BP Global Power. "Going for Grid Parity." BP Global Power, http://www .bp.com/genericarticle.do?categoryId=9013609&contentId=7005395.

Brealey, R. A., S. C. Myers, and F. Allen. *Principles of Corporate Finance*. New York: McGraw-Hill/Irwin Series in Finance, Insurance, and Real Estate, 14 (2005): 561–563.

Bronowski, Jacob. *The Ascent of Man*. Boston: Little, Brown & Co., 1973.

"Budget Outlook 2010." Congressional Budget Office, http://www.cbo.gov/ ftpdocs/108xx/doc10871/BudgetOutlook2010_Jan.cfm.

Campbell-Kelly, Martin and William Aspray. *Computer: A History of the Information Machine.* New York: Westview Press, 2004.

"Canadian Social Credit Movement." Wikipedia, http://en.wikipedia.org/wiki/Canadian_social_credit_movement#Alberta (accessed May 13, 2011).

Cajal, S. R. (1909–11). Translated by Neely Swanson and Larry W. Swanson. *Histology of the Nervous System of Man and Vertebrates.* New York: Oxford University Press, 1995.

Chapman, Shorey H. *The Extortion: A Survival Manual for Hypersensitives in the 21ˢᵗ Century.* Lincoln, NE: iUniverse, 2005.

Clark, J. D. and W. K. Harris. "Fire and its roles in early hominid lifeways." *The African Archaeological Review, 3.* (1985): 3–27.

Clymer, Floyd. *Treasury of Early American Automobiles, 1877–1925.* New York: McGraw-Hill, 1950.

Copeland, B. J. *Colossus: The Secrets of Bletchley Park's Codebreaking Computers.* Oxford: Oxford University Press, 2006.

Daniels, Peter T. and William Bright (eds.). *The World's Writing Systems.* New York: Oxford University Press, Inc., 1996.

"Data Visualizations for a Changing World." World Bank, http://www.google.com/publicdata/ (accessed May 13, 2011).

Davis, M. *Planet of Slums.* New York: Verso, 2006.

DeNavas-Walt, Carmen, Bernadette D. Proctor, and Jessica C. Smith, US Census Bureau.: U. S. Department of Commerce, Economics and Statistics Administration, U. S. Census Bureau.

Current Populations Reports P60–236. *Income, Poverty, and Health Insurance Coverage in the United States: 2008.* U.S. Government Printing Office, Washington, DC, 2009. http://www.census.gov/hhes/www/poverty/poverty08/pov08hi.html.

Dickmanns, Ernst. *Dynamic Vision for Perception and Control of Motion.* London: Springer-Verlag, 2007.

Douglas, C. H. *Economic Democracy (new edition).* Suffork, UK: Bloomfield Books, 1974.

Douglas, C. H. "Credit-Power and Democracy." Accessed May 30, 2011. http://www.archive.org/details/creditpowerdemoc00douguoft.

Energy Information Administration. "Form EIA-861 Data." *Annual Electric Power Industry Report*. Washington, DC DOE/EIA-0348, 2009.

Essinger, James. *Jacquard's Web: How a Hand-Loom Led to the Birth of the Information Age*. Oxford: Oxford University Press, 2004.

Felleman, D. J. and D. C. Van Essen. "Distributed Hierarchical Processing in the Primate Cerebral Cortex." *Cerebral Cortex* 1 (1991): 1–47.

"French high-speed TGV breaks world conventional rail-speed record." Daily Le Parisien, February 14, 2007.

Foner, Philip S. "History of the Labor Movement in the United States: Vol. 1." From *Colonial Times to the Founding of the American Federation of Labor*. New York: International Publishers, 1978.

Gilder, George. *Wealth and Poverty*. New York: Basic Books, 1981.

Gordon, S. H. *Passage to Union — How the Railroads Transformed American Life, 1829–1929*. Chicago: Ivan R. Dee Inc., 1996.

"Grand Challenges for Engineering." National Academy of Engineering, http://www.engineeringchallenges.org/.

Green, M. "Thin-film solar cells: review of materials, technologies and commercial status." *Journal of Materials Science: Materials in Electronics* 18 (October 1, 2009): 15–19.

Grunwald, M. "How the Stimulus Is Changing America." *Time Magazine*, September 6, 2010.

Hallo, William and William Simpson, *The Ancient Near East*. New York: Harcourt, Brace, Jovanovich, 1971.

Harrison, R. "Jeremy Bentham," in Honderich, T. (ed.) *The Oxford Companion to Philosophy*. Oxford: Oxford University Press, 1995.

Hazlitt, Henry. *Economics in One Lesson*. New York: Three Rivers Press, 1988.

Headrick, Daniel R. *Technology: A World History*. New York: Oxford University Press, 2009.

Heide, L. *Punched-Card Systems and the Early Information Explosion, 1880–1945*. Baltimore, MD: Johns Hopkins Press, 2009.

Hodgkin, A. and A. Huxley. "A quantitative description of membrane current and its application to conduction and excitation in nerve." *J. Physiol* 117 (1952): 500–544.

Hogan, Christopher, June Lunney, Jon Gabel, and Joanne Lynn. "Medicare Beneficiaries' Costs of Care in the Last Year of Life." *Health Affairs 20*, no. 4 (July 2001): 188–195.

Hood, Christopher P. *Shinkansen – From Bullet Train to Symbol of Modern Japan*. London: Routledge, 2006.

Hu, Zuliu and Mohsin S. Khan. "Why is China Growing so Fast?" *Economic Issues*, no 8. Washington, DC: International Monetary Fund, 1997.

Hutcheson, D. G. "Moore's Law: The History and Economics of an Observation That Changed the World." *The Electrochemical Society INTERFACE* 14, no. 1. (Spring 2005): 17–21.

Hyman, A. *Charles Babbage, Pioneer of the Computer*. Princeton: Princeton University Press, 1982.

Iagnemma, Karl and Martin Buehler, eds. *Journal of Field Robotics, Special Issue on DARPA Grand Challenge, Part 1*. 12, issue 8. (August 2006): 461–652.

Iagnemma, Karl and Martin Buehler, eds. *Journal of Field Robotics, Special Issue on DARPA Grand Challenge, Part 2*. 23, issue 9. (September 2006): 655–835.

International Energy Agency. *Key World Energy Statistics 2006*. France: Stedi Media, 2006.

Jacobs, O. L. R. *Introduction to Control Theory*. New York: Oxford University Press, 1974.

Jacobson, Mark Z. "Review of Solutions to Global Warming, Air Pollution, and Energy Security." *Energy Environmental Science*. Stanford, CA: Department of Civil and Environmental Engineering, Stanford University, 2008.

Jiasheng, Feng. *The Invention of Gunpowder and Its Spread to the West*. Shanghai, China: Shanghai People's Press, 1954.

Kandel, Eric R., James H. Schwartz, and Thomas M. Jessell. *Essentials of Neuroscience and Behavior*. New York: McGraw-Hill, 1995.

Kelso, L. O. and M. Adler. *The Capitalist Manifesto*. New York: Random House, 1958.

Kelso, L. O. and P. Hetter. *Two Factor Theory: The Economics of Reality*. New York: Random House, 1967.

Kennedy, P. *Rise and Fall of the Great Powers*. New York: Random House, 1987.

Keynes, John M. *General Theory of Employment, Interest, and Money*. Cambridge: Macmillan Cambridge University Press, 1936.

Keynes, John M. *How to Pay for the War: A Radical Plan for the Chancellor of the Exchequer*. London: Macmillan and Co., ltd., 1940.

Kirby, R., S. Withington, A. B. Darling, and F. G. Kilgour. *Engineering History*. New York: Courier Dover Publications, 1990.

Koch, Christof. *The Quest for Consciousness: A Neurobiological Approach*. Englewood, CA: Roberts & Company Publishers, 2004.

Krisher, T. *GM Envisions Driverless Cars on Horizon*. Detroit: Associated Press, January 7, 2008.

Krugman, Paul. "How fast can the economy grow?" *Harvard Business Review*, July/August 1997.

Krugman, Paul. *The Age of Diminished Expectations*. Cambridge, MA: MIT Press, 1994.

Krugman, Paul and Robin Wells. *Macroeconomics,* 2nd ed. New York: Worth Publishers, 2009.

Kurland, Normand G., Dawn K. Brohawn, and Michael D. Greaney. *Capital Homesteading for Every Citizen*. Washington, DC: Economic Justice Media, 2004.

Kurzweil, Ray. *The Singularity Is Near*. New York: Penguin Group, 2005.

Landreth, H. and D. C. Colander. *History of Economic Thought* (4th edition). Boston: Houghton Mifflin, 2002.

Link, Albert N. and John T. Scott. "Public/Private Partnerships: Stimulating Competition in a Dynamic Market." *International Journal of Industrial Organization* 19, issue 5 (April 2001) 763–794.

Mansfield, E. et al., "Social and Private Rates of Return from Industrial Innovations." *The Quarterly Journal of Economics* 91, no. 2 (May 1977): 221–240.

Marinchek, John A. "Will American Manufacturing Jobs be Back?" *American Affairs* (March 23, 2010), http://www.suite101.com/content/will-american-manufacturing-jobs-be-back-a214776.

Marx, Karl. "Critique of the Gotha Programme" (1897). In *Marx/Engels Selected Works, Vol. 3*. Moscow: Progress Publishers (1970), 13-30.

Marx, Karl. *Das Kapital*. Moscow: Progress Publishers, 1887.

"Measuring Global Poverty (2009)." World Bank. http://econ.worldbank.org/WBSITE/EXTERNAL/EXTDEC/EXTRESEARCH/00, content MDK:22452035~pagePK:64165401~piPK:64165026~theSitePK:4693 82~isCURL:Y,00.html.

McCague, James. *Moguls and Iron Men — The Story of the First Transcontinental Railroad*. New York: Harper & Row, 1964.

McCulloch, Warren S. and Walter Pitts. "A logical calculus of the ideas immanent in nervous activity." *Bulletin of Mathematical Biophysics* 5, issue 4. New York: Springer (1943): 115–133.

Moravec, H. *Mind Children: The Future of Robot and Human Intelligence*. Cambridge, MA: Harvard University Press, 1998.

Mundel, E. J. "U.N. Seeks to Curb World's Traffic Deaths." *Washington Post Newspaper*, April 1, 2008.

National Highway Traffic Safety Administration. "National Statistics." Fatality Reporting System, http://www-fars.nhtsa.dot.gov/Main/index.aspx (accessed May 13, 2011).

Niedermeyer E. and F. L. da Silva. *Electroencephalography: Basic Principles, Clinical Applications, and Related Fields*. Baltimore, MD: Lippincot Williams & Wilkins, 2004.

"Overview." Singularity University, http://singularityu.org/about/overview/

Pavlov, I. P. *Conditioned Reflexes: An Investigation of the Physiological Activity of the Cerebral Cortex.* Translated and edited by G. V. Anrep. London: Oxford University Press, 1927.

Pearlstein, S. "Wage cuts hurt, but we need them." *Washington Post Newspaper.* October 13, 2010, A13.

Pigou, A. C., Milton Friedman, and N. Georgescu-Roegen. "Marginal Utility of Money and Elasticities of Demand." *The Quarterly Journal of Economics* 50, no. 3 (May, 1936): 532–533.

Piketty, Thomas and Emmanuel Saez. "Income Inequality in the United States: 1913–1998." *Quarterly Journal of Economics 118,* no. 1 (2003): 1–41.

Pope Leo XIII. *Rerum Novarum, Encyclical on Capital and Labor.* Vatican City: Vatican Publishing House, 1891.

"Prairie Lands Bio-Products Purchases." Prairie Lands, http://www .iowaswitchgrass.com/.

Project Counseling Services, "Deteriorating Bogota: Displacement and War in Urban Centers." *Colombia Regional Report: Bogota.* December 2002.

Raichle, Marcus E. and Mark A. Mintun. "Brain Work and Brain Imaging." *The Annual Review of Neuroscience.* Washington University Library (2006): 449–476.

Rand, Ayn. *Atlas Shrugged.* New York: Penguin, 1957.

Reich, Robert B. "Manufacturing Jobs Are Never Coming Back." *Forbes .com,* May 25, 2009, http://www.forbes.com/2009/05/28/robert-reich-manufacturing-business-economy.html.

Rhodes, Richard. *Dark Sun: The Making of the Hydrogen Bomb.* New York: Simon and Schuster, New York, 1995.

——— *The Making of the Atomic Bomb.* New York: Simon and Schuster, 1986.

Rifkin, Jeremy. *The End of Work.* New York: Penguin Books, 1995.

Rizzolatti, G. and L. Craighero. "The Mirror-Neuron System." *Annu. Rev. Neurosci* 27 (2004): 169—192.

Rojas, Raúl and Ulf Hashagen, eds. *The First Computers: History and Architectures*. Cambridge, MA: MIT Press, 2000.

Samuelson, Paul and William Nordhaus. *Economics, 13th edition*. New York: McGraw-Hill, 1989.

Samuelson, Robert J. "China's stiff upper hand." *Washington Post Newspaper*, November 8, 2010, A15.

Samuelson, Robert J. "The Age of Austerity." *Washington Post Newspaper*, Oct. 11, 2010, A17.

Sarel, Michael. "Growth in East Asia: What We Can and What We Cannot Infer." *Economic Issues*, no. 1. Washington, DC: International Monetary Fund, 1996.

Sauer, Carl O. *Agricultural origins and dispersals*. Cambridge, MA: MIT Press, 1952.

Scott, J. "An Assessment of the Small Business Innovation Research Program in New England: Fast Track Compared with Non-Fast Track Projects," in *The Small Business Innovation Research Program: An Assessment of the Department of Defense Fast Track Initiative*, Edited by Charles W. Wessner, Washington, DC: National Academy Press (2000), 104–140.

Shelley, Mary. *Frankenstein*. Calgary, Canada: Qualitas Publishing, 1818.

Sherman, Arloc and Chad Stone. *Income Gaps Between Very Rich and Everyone Else More Than Tripled In Last Three Decades, New Data Show*. Washington, DC: Center on Budget and Policy Priorities, June 25, 2010.

Shi, Anqing. "How Access to Urban Potable Water and Sewerage Connections Affects Child Mortality." Finance Development Research Group World Bank Working Paper, January 2000.

Singer, P. W. *Wired for War: The Robotics Revolution and Conflict in the 21st Century*. New York: Penguin Press, 2009.

Skinner, B. F. *The Behavior of Organisms: An Experimental Analysis*. Cambridge, MA: B.F. Skinner Foundation, 1938.

Smith, Adam. *The Theory of Moral Sentiment*. Edited by Knud Haakonssen. Boston: Cambridge University Press, 2002.

"Solar Electric Generating Systems." Next Era Energy Resources, http://www.nexteraenergyresources.com/content/where/portfolio/pdf/segs.pdf.

Srodes, James. *Franklin: The Essential Founding Father.* Washington, DC: Regnery Publishing, 2002.

Stephenson, Carl. *Medieval Feudalism.* Ithaca, New York: Cornell University Press, 1942.

US Energy Information Administration. "Table ES1. Summary Statistics for the United States, 1998 through 2009." Electric Power Annual 2009, http://www.eia.doe.gov/cneaf/electricity/epa/epaxlfilees1.pdf.

Taussig, Michael. *Law in a Lawless Land: Diary of a Limpreza in Colombia.* Chicago: The University of Chicago Press, 2003.

Taub, Stephen. "Rich List." *AR Magazine.* April 1, 2010.

"The World Fact Book." Central Intelligence Agency, https://www.cia.gov/library/publications/the-world-factbook/.

"Thrift Savings Plan," https://www.tsp.gov/index.shtml.

Thurow, L. "America among equals," in S. J. Unger's (ed.) *Estrangement: America and the World*, Oxford University Press (1996), New York.

Toth, Nicholas, and Kathy Schick. "Overview of Paleolithic Archeology," in *Handbook of Paleoanthropology Part 3*, edited by Winfried Henke and Ian Tattersall, 1943–1963. Berlin: SpringerLink, 2007.

UN Habitat. "Slums of the World: The Face of Urban Poverty in the New Millennium." Working Paper, 2003.

US Census Bureau. "Poverty." http://www.census.gov/hhes/www/poverty/poverty08/pov08hi.html.

US Department of Energy. "Countries." Energy Information Administration, http://tonto.eia.doe.gov/country/index.cfm (accessed May 13, 2011).

US Department of Energy. "National Renewable Energy Laboratory." http://www.nrel.gov/.

Voltaire. *Candide* (1759). Literature.org, http://www.literature.org/authors/voltaire/candide/.

von Hove, Tann, ed. "More than one billion people call urban slums their home." *City Mayors.* October 5, 2003, http://www.citymayors.com/report/slums.html.

Vonnegut, Kurt. *Player Piano.* New York: Dell Publishing, Random House, 1952.

Waldbaum, Jane C. "From Bronze to Iron: The Transition from the Bronze Age to the Iron Age in the Eastern Mediterranean," in *Studies in Mediterranean Archaeology, LIV.* Göteburg: Paul Astöms Förlag, 1978.

Wiegell, M., T. Reese, D. Tuch, G. Sorensen, and V. Wedeen. "Diffusion Spectrum Imaging of Fiber White Matter Degeneration." *Proc. Intl. Soc. Mag. Reson. Med* 9 (2001): 504.

Worldsteel Association. Steel Statistical Yearbook. Brussels: Worldsteel Committee on Economic Studies, 2010.

World Bank. "World Development Indicators." http://www.google.com/publicdata/explore?ds=d5bncppjof8f9_&ctype=l&strail=false&nselm=h&met_y=ne_gdi_totl_zs&hl=en&dl=en#ctype=l&strail=false&nselm=h&met_y=ne_gdi_totl_zs&scale_y=lin&ind_y=false&rdim=country&idim=country:USA&tdim=true&hl=en&dl=en (accessed May 11, 2011).

World Bank. "World Development Indicators." http://www.google.com/publicdata/explore?ds=d5bncppjof8f9_&ctype=l&strail=false&nselm=h&met_y=ne_gdi_totl_zs&hl=en&dl=en#ctype=l&strail=false&nselm=h&met_y=ne_gdi_totl_zs&scale_y=lin&ind_y=false&rdim=country&idim=country:USA:CHN&tdim=true&hl=en&dl=en (accessed May 11, 2011).

World Bank. "World Development Indicators." http://www.google.com/publicdata/explore?ds=d5bncppjof8f9_&ctype=l&strail=false&nselm=h&met_y=ne_gdi_totl_zs&hl=en&dl=en#ctype=l&strail=false&nselm=h&met_y=ne_gdi_totl_zs&scale_y=lin&ind_y=false&rdim=country&idim=country:USA:CHN:JPN&tdim=true&hl=en&dl=en (accessed May 11, 2011).

World Bank. "World Development Indicators." http://www.google.com/publicdata/explore?ds=d5bncppjof8f9_&ctype=l&strail=false&nselm=h&met_y=ny_gdp_mktp_kd_zg&hl=en&dl=en#ctype=l&strail=false&nselm=h&met_y=ny_gdp_mktp_kd_zg&scale_y=lin&ind_y=false&rdim=country&idim=country:CHN:USA&tdim=true&hl=en&dl=en (accessed May 11, 2011).

World Bank. "Database Directory." http://www.google.com/publicdata/directory.

Wynn, Gerard. *Solar power edges towards boom time.* London: Reuters, October 19, 2007.

Yenne, B. *100 Inventions that Shaped World History.* San Mateo, CA: Bluewood Books, 1993.

INDEX

Page numbers with *italic "n"* or *"nn"* indicates a reference to footnotes
Page numbers with *italic "ill"* indicates a reference to figures, diagrams or tables

reverse engineering human brain and, 94–97
"conditioned reflex," 95
conflict, causes of, 16–17
Congressional Budget Office, on growth of U.S. economy, 44*n*50
Congressional Budget Office (CBO) Data
    compound of estimated rate of growth, 64, 65
    growth in income since 1979, 12*ill*
construction industry, computer controlled robots in, 91
consumer credit, as factor of consumer demand, 21
consumer demand, in market, factors determining, 30
consumer price index, interest rates effect on, 34
consumers, borrowing money by, 34
control theory, on producing instability, 34
cooper tools, development of, 81
Copeland, B. L., 88*n*103
cotton gin, 37, 84
Craighero, L., 96*n*127
credit
    access for investing, 50, 51–52
    formula to prevent inflation when issuing, 57
    impact of explosion of, 32
    infusion into investment of, 71
credit cards, issuing of, 35
Credit-Power and Democracy (Douglas), 48
crystalline structure, knowledge of, 82
Cuba, as totalitarian state, 27
customers
    competition for, 20
    limits in market, 28
Cyrus the Great, 129

D

da Silva, F. L., 97*n*132
Daily Le Parisien, first high-speed passenger train in France, 85*n*92
Daniels, Peter T., 80*n*76, 80*nn*76–77
Darius the Great, 129
Darling, A. B., 84*n*86
DARPA (Defense Advanced Research Projects Agency), 97
Das Kapital (Marx), 25, 25*n*30
Davis, Mike, 14, 14*nn*15–18

De Forest, Lee, 87
Defense Advanced Research Projects Agency (DARPA), 97
Delhi, slum communities in, 14
Democratic Party, doctrine of, 134–135
DeNavas-Walt, Carmen, 11*n*8
depreciation, 66*n*70
design cycles, engineering, economic implications of shortened, 101
design for algae farm, 112–115, 113*ill*, 114*ill*
design systems, computer-aided, 89, 90–91
Dhaka, slum communities in, 14
Dickmanns, Ernst, 92*n*107
diesel power, 85
"difference machine," 88
digital computers, first, 88
dividends
    from investments, 59
    under PIP, 54
    for poor, 73
Domestic Expectations, era of, 134
domestic tranquility, economic growth and, 134–135
Douglas, Clifford Hugh, 48–49
Douglas Credit Party (Oceania Australian), 49
drug trade, as income source, 15

E

Eckert, J. Presper, 88
economic activity, instability of, 31–32
economic cycles, 31–32
Economic Democracy (Douglas), 48
economic establishment, technology and, 100–103
economic futures, 3–4
economic growth
    compound effect of estimated rate of, 65–66
    by creating jobs, 74
    in current capitalist economy, 43–44
    domestic tranquility and, 134–135
    environmental preservation and, 141–142
    in mature economy, 101
    Peoples' Capitalism impact on, 120–121
    since 2008, 28
    solution to declining, 135
    as source of prosperity, 34

supply-side economic theory and fast, 137–138

of U.S., 139

economic justice, world of, 8, 144–145

economic philosophy, arguments around political ideology and, 7

economic policy, under Peoples' Capitalism, 50–51

economic prosperity, jobs and, 38–39

economic strength, military strength and, 125

Economics in One Lesson (Hazlitt), 13, 13n13

economies, focus of communist, 27

economists, Panglossian, 54

economy

    impact of explosion of credit on, 32

    near and long term fix for, 64–68

    Peoples' Capitalism benefits to, 71–75

Edison, Thomas, 87

education, in Muslim slums, 15

EEG (electroencephalograms), 97

Egypt

    advances in technology, 100

    first use of wheel, 83

    hieroglyphs in, 79

electric computers. *see also* computers, development of, 87–89

electric light bulb, 87

electrical card readers, 88

electricity

    cost per kilowatt-hour for electricity, 107

    discovery of, 84, 86–87

    generating power of U.S., 108, 108n147

    powering passenger trains, 85

    waste incinerators generating, 110

electroencephalograms (EEG), 97

electromagnetism, laws of, 86

Employee Stock Ownership Programs (ESOPs), 49

Employee Stock Ownership proposals, 50

encrypted messages, breaking, 88

energy, sources of clean

    about, 105–106

    algae farms, 111–118

    biofuel, 108–111

    solar power, 106–108, 111

energy independence, 128–130

Energy Information Administration data, 107n140, 107n145, 108n146, 116n152

engineering design cycles, economic implications of shortened, 101

Engineering of Mind (Albus and Meystel), 94n117, 148n172

engineering systems, computer-aided, 89

engines, types of internal combustion, 85–86

entrepreneurship, capitalism and, 23

environmental preservation, 141–142

equatorial solar energy, 111

era of austerity, 136–138

era of Domestic Expectations, 134

Essinger, James, 87n99

ethanol, 109

Europe, financial system collapse (2008) in, 28

experiments, economic implications of shortened, 101

exponential growth of economy, 65*ill*

Extortion of "work or die," 74, 74n73

F

Faraday, Michael, 86

farm workers, Industrial Revolution effect on, 36–37, 37*ill*

farming oceans, 111–118, 142

farms

    algae, 111–118

    using computer controlled machines, 91

"father of the computer," 88

Federal Budget, balancing, 65

Federal Reserve

    formula to prevent inflation when issuing credit, 57

    PIP and, 51, 52*ill*

    policy of controlling inflation, 61–62

Felleman, Daniel J,, 96, 96n125

feudalism

    mercantilism and, 21–22

    serfs under, 24

fighter-bombers, pilotless, 122

financial services industry, computers used in, 90

financial system collapse (2008), of Western capitalist countries, 28

fire, domestication of, 79–80

fiscal policies, 32–33

fMRI (functional Magnetic Resonance Imagery), 97

Foner, Philip S., 25n28

food, competing with fuel, 110–112
foreign oil, 128–130
fossil fuel
    burning, 109
    supply of, 105
France, high-speed passenger trains in, 85
Frankenstein (Shelley), 102, 102n137
Franklin, Benjamin, 86
free market capitalism. *see also* capitalism;
    Peoples' Capitalism
    about defects of, 27–28
    capacity of, 17
    competition for customers in
        marketplace, 20
    fiscal and monetary policies, 32–34
    globalization and, 39–42
    high-tech industry creating jobs, 42–43
    history of, 21–23
    instability of economic activity, 31–32
    limited access to credit for investment,
        35
    on modernization of China, 40
    owners vs. workers, 29–30
    under Peoples' Capitalism, 73
    persistence of poverty and, 28–29
    principle factor of production, 102
    principle of private ownership, 19–20
    profit motive in, 20
    purpose of, 24
    responding to market demand, 20–21
    slow economic growth, 43–44
    tendency toward monopoly, 30–31
    unemployment and, 35–39
    workers under, 23–25
free trade, job losses and, 72–73
Friedman, Milton, 33, 33n36
fuel, competing with food, 110–112
functional Magnetic Resonance Imagery
    (fMRI), 97

G

Gabel, Jon, 53n55
Galvani, Luigi, 86
gangs
    sophistication of, 15
    violent crimes of, 16
"garbage mountain" slum of Navarro, 14
gas, burning, 109
gas turbine engines, 85
GDP (Gross Domestic Product)

between 1939 and 1945, 43
    capital deepening and, 58, 58n65
    consumer spending and, 21
    estimate government spending as percent
        of, 65
    exponential growth of economy, 64, 65*ill*
    growth since 2008, 28
    impact of increasing investment rate on,
        120
    increase and computer controlled
        machines, 99
    during Reagan and Bush (G. W.) years,
        136n164
    US capital formation rate (2007) and,
        57, 57n64
    U.S. investment and, 54–55
*General Theory of Employment, Interest, and
    Money* (Keynes), 63, 63n68
geothermal energy, 106–108
Germany
    Enigma encryption machines, 88
    Military University, in Munich, 91–92
    production of automobile in, 85–86
Gerson, Michael, 136
Gilded Age, 31
Gilder, George, 135n163
Global Hawk (UAV), 122
global warming, 105
globalization, 39–42, 135
goods and services, production of, as
    machine intelligence grows, 3
Google, 93
Gordon, S. H., 85n89
government debt, solution to, 135
government spending
    as investment, 53
    rate of, 65
    WW II and, 33
GPUs (Computer Graphics Processing
    Units), 94
Great Britain
    effect of Keynesian stimulus programs,
        136
    railroads in, 84
    Social Credit movements in, 49
Great Depression, 33
Greece
    advances in technology, 100
    social upheaval in, 136
Green, M., 106n139

growth of economy, exponential, 65*ill*
Grumman Aerospace Corporation, pilotless
  fighter-bombers developed by, 122
Grunwald, M., 107*n*143
gunpowder, invention of, 82

H
Hallo, William, 81*n*79
Hammurabi, 129
Harris, W. K., 79*n*75
Harrison, R., 12*n*10
Harvard Business Review, on speed of
  economic growth, 136–137, 137*n*165
Hashagen, Ulf, 88*n*104
Hazlitt, Henry, 13, 13*n*13
Headrick, Daniel R., 84*n*88
heat, using biofuels for, 110
hedge fund managers, incomes during
  recession of 2009, 11, 11*n*9
Heide, L., 88*n*102
Hertz, Heinrich, 86
high-speed passenger trains, 85
high-tech industry jobs, 42
Ho Chi Min, 27
Hodgkin, Alan, 95, 95*n*124
Hogan, Christopher, 53–54
Hollerith, Herman, 88
Hong Kong, economic growth of, 43–44
Hood, Christopher P., 85*n*91
"How Fast Can the Economy Grow?"
  (Krugman), 136–137, 137*n*165
How to Pay for the War (Keynes), 63, 63*n*67
Hubel, David, 95–96
human brain, computers and, 42–43
human brain-power, computers replacing,
  90
human-level machine intelligence, for
  military applications, 121–124
hunter-gatherer groups, 80–81
Hutcheson, D. G., 93*n*112
Huxley, Andrew, 95, 95*n*124
hybrid automobiles, 109
hydroelectric power, 105–106
hydrogen bomb, 125
Hyman, A., 88*n*100

I
I Robot (movie), 102
IBM, founding of, 88

ideal world
  reality of, 10–17
  vision of, 8–10
idealistic youth, in slum communities, 16
IEDs (Improvised Explosive Devices), 121,
  123
Improvised Explosive Devices (IEDs), 121,
  123
income, disparity in wealth and, 11–17
income sources, for extremists, 15
incomes
  capital assets and, 53
  disparity in U.S. since 1979, 12*ill*
  independent of work, 143
  per capita annual income floor, 66–67,
    67*ill*, 69–70, 69*ill*
India
  globalization and, 40
  producing steel, 82
  slum dwellers in, 10
Industrial Revolution
  disparity in wealth and income during,
    11
  effect of productivity growth generated
    by, 36–37, 37*ill*
  entrepreneurship and, 23
  invention of steam engine, 83–85
  justification in disparity in wealth since,
    13
  technology of, 1–2
  wheel as central component of, 83
industries, focus of communist, 27
inflation
  controlling, 33–34, 61–64
  formula when issuing credit to prevent,
    57
  problem of, 59–68
  unsustainable economic growth and, 44
inflationary gap, 63
Information Sciences, on computation and
  representation in brain, 95*n*118, 96*n*129
information sources, in slum communities,
  15
information technology, predictions about,
  101–102
information technology R&D projects,
  companies investing in, 56*n*62
information technology revolution
  basis of, 2–3
  developing economic system in, 4–5

Krisher, T., 86n95
Krugman, P., 43n43
Krugman, Paul, 134, 134n162, 136–137, 137n165
Kurland, Norman, 50, 51, 51n53
Kurzweil, Ray, 43, 43n42, 93, 93n113, 148, 148n171

L

labor, profits vs., 30
labor market, under capitalism, 24
labor organizations, influence on wages, 13
labor shortages, influence on wages, 13
labor theory of value, classical economic theory and, 3
laissez-faire capitalism, 23
Landreth, H., 22n25
language, development of techology and, 80
laws of electromagnetism, 86
Layard, Richard, 2n3
League of Rights (Oceania Australian), 49
Lenin, Vladimir I., 16, 27
Leo XIII, Pope, 25, 25n29
Les Démocrates (Canada), 49
Link, Albert N., 56, 56n62
Luddite riots (England), in 1811 and 1812, 41, 84
Lugaresi, E., 86n97
Lunney, June, 53n55
Lynn, Joanne, 53n55

M

machine intelligence. *see also* computer controlled machines
    material wealth and, 43
    for military applications, 121–124
machine tools, development of, 84
machines, modern manufacturing, 3
mandatory savings plan, 62–63
Manitoba Social Credit Party, 49
Mansfield, Edwin, 55, 55n61, 57, 71
manufacturing
    introduction of technology into, 38*ill*
    jobs required for manufactured goods since, 38, 38*ill*
    under mercantilism, 22
"Manufacturing Jobs Are Never Coming Back" (Reich), 38, 41, 41n41
Mao Zedong, 16, 27

Marconi, Guglielmo, 86–87
Marinchek, John A., 40n38
market
    competition for customers, 20
    limits on customers in, 28
    responding to demand of, 20–21
market crashes, frequency of, 31
market demand, determining of, 30
market supply and demand for labor, 13
Marx, Karl, 25, 25n30, 26, 26n33
Marxism, appeal of, 27
Mathematical Biosciences, on cerebellar function, 97n130
Mauchly, John William, 88
Maxwell, James, 86
McCague, James, 85n90
McCormick reaper, 84
McCulloch, Warren, 95, 95n123
means of production
    distribution of ownership of, 3–4
    owners of, 13
Medicare and Medicaid
    cut spending in, 120
    estimate of percent of GDP for, 65
    expenditures, 53, 53n55
    solving problem of, 71
    solving problems of, 64
mercantilism
    about, 22–23
    capitalism and, 21
    workers under, 24
Meystel, Alexander M., 94n117, 148n172
middle class
    benefits of Peoples' Capitalism for, 73
    emergence of, 2, 13
    under Peoples' Capitalism, 133
    primary source of income, 102
    reversing decline of, 4
Middle East
    grievances against West, 129
    revolutionary movements in, 26
    warfare in, 17
military overstretch, consequences of, 119–120
military strength, economic strength and, 125
Military University, in Munich, 91–92
Miller, Matt, 65
mine employees, 24
Mintun, Mark A., 97n131

"A Model of Computation and Representation in the Brain" (Albus), 95n118, 96n129, 148n173
model T Ford, 85–86
modernization, wages and, 40–41
monetary policies, 33, 34, 44
monetary restraint, controlling inflation through, 61–62
monks, working for common good, 26
monopolies, capitalism tendency toward, 30–31
Moore, Thomas, 26
Moore's Law, 93, 94
Moravec, H., 94n116
motor vehicles. *see* automobiles
movies, as information source in slum communities, 15
Mumbai, slum communities in, 14
Mundel, E. J., 92n111
Muslim slums, education in, 15
Muslims, intolerance for infidels, 129, 129n160
mutual funds
   investing in, 144–145
   PIP and, 51–52, 52*ill*

N

nanotechnology, 82
NASA Ames Research Center, 93
National Dividend, about, 48–49
National Highway Traffic Safety Administration Fatality Reporting System, 92n110
National Renewable Energy Laboratory (NREL), 111, 111n151
National Science Foundation (NSF), 97
National Security Agency (NSA), 97
national security imperative, 121–124
national states, development of, 22–23
natural gas, supply of, 105
Nebuchadnezzar, 129
neurons
   in brain, 95
   in central nervous system, 96
neuroscience, understanding computational mechanisms, 95–98
New Capitalist Manifesto (Kelso and Adler), 49
New Democracy (Canada), 49

New Zealand Democratic Party for Social Credit, 49
Newcomen, Thomas, 83–84
Next Era Energy Resources brochure, on solar electric generating systems, 107n144
Niedermeyer, E., 97n132
North Korea, as totalitarian state, 27
Northern Ireland, Social Credit movements in, 49
NREL (National Renewable Energy Laboratory), 111, 111n151
NSA (National Security Agency), 97
NSF (National Science Foundation), 97
nuns, working for common good, 26

O

Obama's Stimulus Package, 107
Oceania Australian, 49
oceans, farming, 111–118, 142
off-shoring of production, 72–73
Office of Naval Research (ONR), 97
oil
   burning, 109
   buying foreign, 128–130
   supply of, 105
ONR (Office of Naval Research), 97
"operant conditioning," 95
opportunity, world of prosperity and, 8
original sin, Christian doctrine of, 10
Ottoman empire, 129
Ottumwa Generating Station, in Iowa, 110, 110n148
owners vs. workers, 29–30

P

Pakistan, Predator drones in, 121–122, 129
Parmeggiani. P. L., 86n97
passenger trains, high-speed, 85
Paul VI, Pope, on development of peoples, 131
Pavlov, Ivan, 95, 95n121
Pearlstein, Steven, 137n167
Peoples' Capitalism. *see also* capitalism; free market capitalism
   about, 47–51, 47n51
   access to credit for investing, 51–52
   addressing needs and wants of people, 133
   benefits to economy, 71–75, 132

productivity growth, worker-owners and, 74
professional societies, purpose of, 24
profits, from industry, distributing, 48
profits vs. labor, 30
program computers, stored, 88
Project Counseling Services, 14n19
prosperity, world of opportunity and, 8, 17
Protestant Reformation, 23
PSP (Personal Savings Plan), 50, 58–59, 62–64, 72
PSP (Personal Savings Program), about, 4–5
psychology, understanding computational mechanisms, 95
punched cards
  "analytic machine" using, 88
  controlling automatic loom, 87–88
  mechanical tabulators based on, 88

Q

al Qaeda, 129

R

R&D projects, companies investing in information technology, 56n62
radar, development of, 87
radios, commercial, 87
Raichle, Marcus E., 97n131
railroad, first public steam, 84–85
railroad construction crews, 24
Ralliement créditiste (Canada), 49
Ralliement créditiste du Québec, 49
Rand, Ayn, 13n12
rate of investment
  of China, 55, 55n58, 57, 128, 128n158
  of United States (U.S.), 54–55, 127–128, 127n157, 128n158
"RCS: A Cognitive Architecture for Intelligent Multi-Agent Systems" (Albus and Barbera), 91n106
Reagan, Ronald, 135, 136, 145
real estate loans, 35
reality of ideal world, 10–17
recession, unemployment and, 35–39
recession of 2009, hedge fund managers incomes during, 11
Recovery Act's Four Investment Goals, 107
Reese, T., 97n133
reference model architectures, evolving of, 91

Reformation, Protestant, 23
Reich, Robert B., 40n39, 41, 41n41
religious establishment, use in society of, 13
religious extremists, recruits for, 15
religious intolerance, 129, 129n160
religious tradition, pessimism in, 10
Renaissance, trade during, 22
Republican Party, doctrine of, 134
research and development
  funding for, 72
  government spending on, 47, 137
  impact of, 60
  investment in, 100, 101, 136, 148
  as investments, 53
  public support for, 72
resistance movements, organizing, 16
resource allocation, under communism, 27
returns to capital, 56
reverse engineering, economic implications of, 98–99
"Reverse Engineering the Brain" (Albus), 94n115
rich people
  demographics of very, 13
  in Gilded Age, 31
  investment banking and, 35
  physical violence and, 16
  reason for working, 74
Rifkin, Jeremy, 142–143, 143n170
"Rights and Duties of Capital and Labor" (Leo XIII, encyclical), 25, 25n29
The Rise and Fall of the Great Powers (Kennedy), 119, 119n153, 125
Rizzolatti, G., 96n127
roadside bombs, 121
robots
  computer controlled, 91
  economic implications of, 98–99
  filling jobs, 145
  in weapons systems, 89–90, 122–124
Rocket (locomotive), 84
rocket engines, 85
Rojas, Raul, 88n104
Romania, social upheaval in, 136
Romans
  advances in technology, 100
  weapons used by, 82
Roosevelt, Franklin, 33
Roosevelt, Teddy, 31
Russia, oil fields in, 128

S

Saez, Emmanuel, 12*n*11
Samuelson, Robert J., 40*n*40, 137, 137*n*166
Santiago de Cali, Columbia, 14
satellite communication, development of, 87
Saudi Arabia, U.S. troops in, 129
saving and thrift, social customs of, 23
savings, mandatory, 62–63
SBIR (Small Business Innovation Research) Program, R&D projects, 56, 56*n*63
Schick, Kathy, 79*n*74
school of behaviorism, 95
Schwartz, James H., 95*n*119, 96*n*128
science, about, 77–78
Scott, J., 56*n*63, 57
Scott, John T., 56, 56*n*62
seaweed, 111–112
Securities and Exchange Commission, PIP and, 52
Shakers, 26
Shakespeare, R., 50, 152
shareholder return on investment, 66*n*70
Shelley, Mary, 102, 102*n*137
Sherman, Arloc, 134*n*161
Shi, Anging, 15*n*21
Shinkansen "Bullet Train," 85
Simpson, William, 81*n*79
Singapore, economic growth of, 43–44
Singer, P.W., 10*n*6, 15*n*22, 98, 124, 124*n*154
Singularity University, 93, 93*n*114
Skinner, B. F., 95, 95*n*122
slum communities
    capitalism and, 29
    children in, 16
    information sources in, 15
    living conditions in, 10–11
    social and economic demographics of, 14
    in South Asia, 14
    war in urban slums, 15
Small Business Innovation Research (SBIR) Program, R&D projects, 56, 56*n*63
Smith, Adam, 28, 28*n*34
Smith, Jessica C., 11*n*8
social and economic demographics
    of middle class, 12–13
    of slums, 14
social benefits
    of investment, 60–61

of Peoples' Capitalism, 71
Social Credit, economic reform movement, 48
Social Credit Board (Canada), 49
Social Credit Party (New Zealand), 49
Social Credit Party of Alberta, 49
Social Credit Party of Canada, 49
Social Credit Party of Great Britain and Northern Ireland, 49
Social Credit Party of Ontario, 49
Social Credit Party of Saskatchewan, 49
Social Credit Party (Solomon Islands), 49
social return, as defined in SBIR R&D projects, 56
Social Security
    cut spending in, 120
    estimate of percent of GDP for, 65
    solving problem of, 64, 71
society, use of political power in, 13
software development, evolving of, 91
solar power, 106–108, 111
Solomon Islands, 49
Somalia, Predator drones in, 121–122
Sorensen, G., 97*n*133
South America, revolutionary movements in, 26
South Asia (Karachi, Mumbai, Delhi, Kolkata, and Dhaka), slum communities in, 14
South Korea
    economic growth of, 43–44, 139
    poor finding jobs in, 40
Southeast Asia, revolutionary movements in, 26
Soviet Union
    first atomic bomb, 125
    as totalitarian state, 27
Spain, social upheaval in, 136
Srodes, James, 86*n*96
Stalin, Joseph, 27
Stanford University, fully autonomous vehicles demonstrated at, 92
Star Wars (movie), 102
steam engine, invention of, 83–84
steel, process of making, 81–82
steel plow, invention of, 84
Steel Statistical Yearbook, 2010, 82*n*82
Stephenson, Carl, 21*n*24
Stephenson, George, 84
Stephenson, Robert, 84

Wiegell, W., 97*n*133
Wiesel, Torsten, 95–96
wind power, 106
Wired for War (Singer), 98, 124, 124*n*154
wireless telegraph, invention of, 86–87
Withington, A. B., 87*n*98
work, life without, 142–143
work ethics of middle class, 13–14
worker-owners, productivity growth and, 74
workers
    in economic systems, 23–25
    owners vs., 29–30
    under Peoples' Capitalism, 133
working, by choice, 74
World Bank
    capital formation rate of Japan, 55*n*59
    on GDP during Reagan and Bush (G. W.) years, 136*n*164
    rate of economic growth of U.S. and China, 128*n*159
    on rate of investment of China, 55*n*58
World bank, on rate of investment of U.S., 127*n*157
World Bank
    on rate of investment of U.S. and China, 128*n*158
    on urban poverty, 14–15
    on U.S. economic growth, 139
    on U.S. investment, 44*n*47
    on U.S. investment and GDP, 54*n*57
    "*World Development Indicators,*" 10, 10*n*5
"*World Development Indicators*" (World Bank), 10, 10*n*5
world peace, stability and, 144–145
World War II (WW II)
    breaking encrypted messages, 88
    government spending and, 33
    jobs required for manufactured goods since, 38, 38*ill*
    mandatory savings proposed at beginning of, 63
    railroads after, 85
    on rate of investment of U.S. since, 127–128
    returns to capital during, 56
    U.S. average annual growth since, 44
    U.S. economic growth before, 139
writing, earliest form of, 79
Wynn, Gerald, 107*n*141

Y

Yenne, B., 83*n*84
youth
    in slum communities, 15, 16